Garbage in the Garden State

Ceres
RUTGERS
STUDIES
IN HISTORY

Lucia McMahon and Christopher T. Fisher, Series Editors

New Jersey holds a unique place in the American story. One of the thirteen colonies in British North America and the original states of the United States, New Jersey plays a central, yet underappreciated, place in America's economic, political, and social development. New Jersey's axial position as the nation's financial, intellectual, and political corridor has become something of a signature, evident in quips about the Turnpike and punchlines that end with its many exits. Yet, New Jersey is more than a crossroad or an interstitial *elsewhere*. Far from being ancillary to the nation, New Jersey is an axis around which America's story has turned, and within its borders gather a rich collection of ideas, innovations, people, and politics. The region's historical development makes it a microcosm of the challenges and possibilities of the nation, and it also reflects the complexities of the modern, cosmopolitan world. Yet, far too little of the literature recognizes New Jersey's significance to the national story, and despite promising scholarship done at the local level, New Jersey history often remains hidden in plain sight.

Ceres books represent new, rigorously peer-reviewed scholarship on New Jersey and the surrounding region. Named for the Roman goddess of prosperity portrayed on the New Jersey State Seal, Ceres provides a platform for cultivating and disseminating the next generation of scholarship. It features the work of both established historians and a new generation of scholars across disciplines. Ceres aims to be field-shaping, providing a home for the newest and best empirical, archival, and theoretical work on the region's past. We are also dedicated to fostering diverse and inclusive scholarship and hope to feature works addressing issues of social justice and activism.

Jordan P. Howell, *Garbage in the Garden State*
Maxine N. Lurie, *Taking Sides in Revolutionary New Jersey: Caught in the Crossfire*
Jean R. Soderlund, *Separate Paths: Lenapes and Colonists in West New Jersey*

Garbage in the Garden State

••••••••••••••••••••••••••••••••

JORDAN P. HOWELL

Rutgers University Press
New Brunswick, Camden, and Newark, New Jersey
London and Oxford

Rutgers University Press is a department of Rutgers, The State University of New Jersey, one of the leading public research universities in the nation. By publishing worldwide, it furthers the University's mission of dedication to excellence in teaching, scholarship, research, and clinical care.

Library of Congress Cataloging-in-Publication Data
Names: Howell, Jordan P., author.
Title: Garbage in the Garden State / Jordan P. Howell.
Description: New Brunswick : Rutgers University Press, [2023] |
    Series: Ceres: Rutgers studies in history | Includes bibliographical references and index.
Identifiers: LCCN 2022029237 | ISBN 9781978833395 (paperback) |
    ISBN 9781978833401 (hardback) | ISBN 9781978833418 (epub) |
    ISBN 9781978833432 (pdf)
Subjects: LCSH: Refuse and refuse disposal—New Jersey.
Classification: LCC TD788.4.N5 H69 2023 | DDC 628.4/409749—dc23/eng/20221011
LC record available at https://lccn.loc.gov/2022029237

A British Cataloging-in-Publication record for this book is available from the British Library.

References to internet websites (URLs) were accurate at the time of writing. Neither the author nor Rutgers University Press is responsible for URLs that may have expired or changed since the manuscript was prepared.

rutgersuniversitypress.org

For everyone who used to, does, or will live in New Jersey—the once and forever Garden State

Contents

Garbage in the Garden State

1

Introduction

•••••••••••••••••••••

TONY SOPRANO: [voice-over, to therapist] Next, I had a breakfast meeting.
   I was called in to consult by a garbage hauling company I represent.
   [to Sal Bonpensiero] Alright, what's the story with Triboro Towers?
SAL BONPENSIERO: Well, the site manager wants to renew his contract with
   Dick, but this Kolar Sanitation . . .
DICK BARONE: [interrupts, agitated] It's another nationwide company.
SAL: Yeah, the Kolar brothers . . . They'll haul paper, plastic, AND the
   aluminum for $7,000 a month less than Dick . . .
TONY: [resigned, disappointed] The f-king garbage business . . .
SAL: Yeah I know . . . It's all changing.[1]

Popular depictions of New Jersey like the one above offer some insights as to why
waste management looms large in the popular imagination of the state. In the
critically acclaimed television series *The Sopranos*—set in New Jersey—the main
character, Tony Soprano, ostensibly owns and operates a trash collection company (among other things). For whatever problems viewers may have with *The
Sopranos'* depictions of Italian-Americans, mental health, or familial relationships, that New Jersey is a place where the garbage disposal business is a prominent aspect of daily life is taken as a given. *The Sopranos* is hardly alone in this
characterization. The animated comedy *Futurama* describes New Jersey as a literal wasteland. In one episode, the narrator offers a simple explanation for New
York City's scramble to find a waste disposal solution: "New Jersey was full."[2]

For many, the picture presented by fictional works like *The Sopranos* or
Futurama can be easily verified by even the most superficial, limited personal

1

experiences. Consider, for example, the sights seen by many folks driving along the state's busiest roads, like the New Jersey Turnpike.[3] From the car, the physical environment of the state is characterized by landfills and other disposal facilities, apparently confirming the perception that New Jersey is primarily a dumping ground for New York, Philadelphia, and summer tourists at the shore. There is a perception that trash is ubiquitous, endemic, and weird in the Garden State. An article in *The Star-Ledger*, a major newspaper for the state, reported in 2011 that: "Nearly a half-million pieces of debris ranging from plastic cup lids to toilet seat lids were left on New Jersey beaches last year . . . A bag of heroin, a 10-gallon gas tank, five pairs of underwear, a duck caller and a plastic cow were among the nearly half-million pieces of trash picked up from New Jersey's beaches by volunteers last year. It may appear that everything but the kitchen sink turned up. But that is wrong: There was one of those, too."[4] Proud New Jerseyans, however, like to combat the perception that the state is little more than a trash heap. Far from being a dump site, in fact over 20 percent of the state is preserved or protected land. There are also some very special, well-managed ecosystems like the Pinelands National Reserve that are truly unique in the Western Hemisphere. One of the largest contributors to the state economy is tourism centered on beach and ocean ecosystems; these require careful environmental monitoring. Ballot initiatives and bond measures aimed at supporting environmental, agricultural, and conservation issues typically receive strong public support. So despite what you might see from your car window on the Turnpike, New Jersey is in many ways an environmental champion.

The realities of waste management—and many environmental issues—in New Jersey are actually between these extremes of dumping ground and eco-paradise. It is absolutely true that for decades New Jersey accepted countless tons of waste materials from New York and Philadelphia, and that tourism, along with many other industries in the state, has left behind what can only be described as a landscape of waste disposal. Most New Jersey towns (some 565) did have their own dumps, and some had several. That means, there really were landfills all over the state, many of them operated with little oversight and placed in environmentally sensitive locations like wetlands and swamps. And yes, organized crime did play a significant role in the history of waste management here. It is equally true, however, that New Jersey has quite progressive waste management laws and targets, especially within the U.S. context. Public officials, activists, and businesses have dedicated extensive time and effort (and money) to designing and operating comprehensive systems of waste collection and disposal. Recycling, composting, and incineration with energy recovery ("waste-to-energy," or WTE) facilities are alive and well in New Jersey and operate at very high environmental standards, despite a growing trend toward simply exporting the state's trash to landfills out of state. Since the 1960s, virtually every approach to managing waste has been studied and implemented among New Jersey's numerous towns and counties, making it a living laboratory for waste management policy. In short, while the state might lag behind in some areas, it is an innovator in others. New Jersey really has seen it all as far as waste management is concerned.

Garbage in the Garden State explores the history of waste management as it unfolded in New Jersey from the late nineteenth century to the early twenty-first century, in order to understand how we arrived at the system we have and identify some areas in which it could be improved. The analysis of the specific case of New Jersey is useful given the state's dynamism and challenging demographic, ecological, and economic contradictions. These are complicated further given that *waste* itself is a unique environmental challenge subject to shifting public perceptions and disagreement over definitions. For this reason, I argue that a one-size-fits-all approach to solving waste management problems is unlikely to be effective; by the same token, a fragmented and uncoordinated system of waste management has led to unsatisfactory economic and environmental outcomes. As this analysis of New Jersey's waste management infrastructure shows, our perception of waste plays an important role in how we treat it: seeing waste primarily as an environmental threat will lead us to one set of outcomes, but approaching waste as a transformative economic opportunity might lead us to another—perhaps with even stronger ecological performance.

Is This Really a Book about Garbage? In New Jersey?

Why study trash at all, let alone in the limited context of a state like New Jersey? There are a few good reasons.

First, the New Jersey experience is quintessentially American. Developments in this state often foreshadowed changes that subsequently unfolded elsewhere in the country. We need only to examine the economic history of New Jersey to see how this is true: New Jersey transitioned from an agricultural to an industrial to a service and knowledge economy in ways that would be repeated across the United States. These transitions included many of the shifts in demographics, ways of life (and especially, suburbanization), patterns of work, and environmental management that have come to define life in twenty-first century America.[5] New Jersey is at the heart of a demographic megaregion stretching between Richmond and Boston, a region accounting for almost 20 percent of all Americans. New Jersey alone is home to about nine million people and over half-a-trillion dollar economy, crammed into the fourth-smallest state in terms of geographic area.[6] New Jersey has always been wedged between the major political and economic centers of New York and Philadelphia, and while it is certainly true that the economy and culture of the state have evolved in large part as a response to those markets, a distinctive New Jersey economy and culture have also been established in response to local environments and ways of life.[7] On top of that, the state is a study in contradictions: New Jersey encompasses both extreme wealth and shocking poverty; dense urban neighborhoods lie within minutes of vigorously protected open space; a powerful state government and 565 individual municipalities tussle endlessly over taxes, civil rights, and schools; all while heavy industry, demand for housing, and a valuable tourist trade threaten some of the most fragile and beautiful ecosystems in the Western Hemisphere. This diversity and dynamism lead to both gridlock and impressive political compromise, not least surrounding management of the environment and the question of how to handle the seemingly endless flow of waste materials produced by New Jerseyans and their businesses. This book examines New Jersey's responses to the waste management puzzle because understanding how solutions have been developed and implemented (or not!) in this most contradictory of places offers insights into the environmental policy-making process that might be useful to New Jerseyans and residents of other states alike, as well as politicians, environmentalists, scholars, and researchers anywhere.

Second, waste management is itself a fascinating, special type of environmental issue, because whether something is *waste* or not is frequently a question of perspective more than incontrovertible fact. "Is this thing emitting radioactive particles, or not?" is a much different question from "is this thing waste, or not?" Materials included in the category of waste can be subjective in a way that many other types of pollution may not be, and definitions can change from person to person. Policymakers, business owners and operators, private citizens, and environmental activist groups might each understand waste as a different thing, or set of things, along a continuum of materials, rather than as a definitive single substance. These shifts in perspective impact our thinking on what to do about waste, or even if we should do anything at all. What is perhaps most interesting about this phenomenon is tracing how definitions of waste have changed as New Jersey communities' fortunes have waxed and waned over time. Economic, demographic, and urban development changes have made impacts on the types of materials that are collected and disposed of (and how this work is done). Important shifts in attitudes toward waste have transpired over the decades, even as the state government in New Jersey came to statutorily classify certain materials as waste. Where previously, waste was perceived as an aesthetic nuisance, economic burden, or ecological threat, over time—and not always with the purest of intentions—waste in New Jersey has also been described as a potential energy source, feedstock for ailing agricultural soils, lifeline for municipal budgets, and of course, locus for organized crime. This book dissects the seemingly monolithic challenge of waste management into its component parts, and examines how stakeholders in New Jersey cooperated and competed with one another to define both the problem of waste management and its possible solutions. The policy and infrastructural implications of competing understandings of waste are highlighted, as well as the related impacts on both environmental conditions and the expenditure of vast sums of public and private money.

Third, examination of waste management issues is a front-row seat for witnessing just how intertwined environmental problems are with social, political, engineering, and economic concerns. Waste management decisions insert themselves into every aspect of our lives, every day, even when we are not explicitly aware of their presence. Consider even a mundane aspect of managing personal appearance—shaving with a disposable razor. You may finish your shaving and think, "The components of this disposable razor must be recyclable—it's plastic and steel. I'll do the right thing for the Earth and drop it in the recycling bin." But it turns out to be far more complex:

1   What type of plastic is the razor handle? The plastic may be acceptable for recycling in one place, but not another. Despite the fact that the material itself does not change depending on your location, one town might consider the plastic recyclable while another does not. This decision is probably based on the cost of collecting and sorting different materials, and the anticipated amount of money a town could get in return for selling this particular type of plastic (which is probably not much to begin with). Does it make sense for your town to spend money on collecting materials with a low resale value, or even take a loss on recycling, when other matters (like funding the school system) may be more pressing? In a world of limited public finances, why not just send the razor to the least-expensive disposal option— probably a landfill? Why does not someone (government?) require that all plastic in disposable razor handles be of the same type, and be easily recyclable? Or maybe, do something to support the market for recycled plastics so that they are worth more?

2   How will workers separate the blade from the plastic? Should *you* do it to save them the time? Can you separate the blade safely? If so, is there even enough metal to be reused? Could your intention to pursue what seems like an ecofriendly disposal method actually endanger the workers at the processing facility, who might be cut by the used blade? What types of people work at a recycling facility anyway, and do they have good insurance if they get badly cut?

3   Is recycling this disposable razor even *worth it* from an energy and materials perspective? That is to say, would reusing the plastic and steel from this item save more energy and raw materials than making a new

disposable razor from scratch? Or does processing the disposable razor actually use *more* energy overall? Maybe it is better if the razor goes into the waste-to-energy incinerator and is burned up? At least then the petrochemicals in the plastic handle are contributing to producing electricity and steam. But does encouraging incineration detract from other *ecofriendly* sources of electricity, like wind and solar? Is waste as a fuel source really a step toward environmental sustainability?

4 Maybe you should invest in a reusable, single-blade shaving handle? You just replace the blade, but keep the same handle. This would seem to generate less waste. Still, there is the issue of the used blades. And what if you fly frequently? In the United States at least, your blades will probably be confiscated at airport security, so you are back to a single-use disposable razor again. Should Transportation Security Administration (TSA) make security exceptions in the name of reducing waste? And what if you cut yourself more frequently with the reusable equipment? Are your shaving preferences and equipment choices significant enough environmentally to make you accept an inferior shaving experience? If you do not change your behavior because of ecological concerns, what does that say about you as a human being and environmental citizen?

What transforms this type of personal eco-anguish into a substantial policy issue is the scale at which it is repeated every single day. We can multiply this individual line of questioning by the billions of disposable razors used for whatever purpose every year; then multiply again by the countless other products and materials we come in contact with in the course of a given day, for which we must decide what to do when we are finished with them. Then multiply once more by all of the businesses that produce waste materials, at even greater rates than we do as private individuals, and for which waste represents not only unwanted materials but also a loss of time, energy, and money. Eventually, we reach a point where the amount of unwanted material is so great that we insist on some sort of public action to create or manage a system for collecting and disposing of it all. We also reach a point where we realize that collection and disposal technologies alone have ceased to be completely effective, and whether for economic, moral, or ecological reasons, changes to our behaviors as private citizens and businesses are necessary in order to avoid drowning in refuse. These realizations are the intellectual motive behind every single waste management policy ever crafted, in all of history.

This book is not a philosophical contemplation of waste, or what it means to create waste.[8] However, philosophical issues are never far from the surface when examining waste management systems. Our choices about what we throw away, recycle, compost, or incinerate; where we locate landfills and processing facilities; how we pay for waste management and the extent to which we want government involved in the process; the types of materials we think should be incorporated into buildings and manufactured products; all of these have roots (however subtle) in our personal relationships to the physical environments we rely on for survival. Our definitions of waste say something about the types of homes and communities we want to live in. From that perspective, the first argument in *Garbage in the Garden State* is that we must avoid one-size-fits-all solutions to managing waste materials. That is to say, we as taxpayers, as collection and disposal customers, and as people who produce waste, should view with suspicion assertions (from any source) proclaiming that "incineration is good/bad" or "recycling is good/bad" or "landfills are un/necessary" or that "sustainable waste management means doing X, Y, or Z." Effective waste management has to be context-specific, and this book highlights how New Jersey, and regions within New Jersey, have developed their own configurations of waste management policy and infrastructure that stakeholders find acceptable, reliable, and maybe even sustainable and resilient. These will—as they must—look different from the arrangements devised in other places and at other times. There is no one universally correct way to handle waste management.

With that said, an analysis of various waste management infrastructures quickly reveals that many are inefficient both economically and ecologically. One reason why, as the second argument in this book asserts, is that waste management systems are exceedingly fragmented. Until the 1960s, garbage was an issue dealt with almost exclusively by individual cities and towns, and it was only with the rise of a broader environmental consciousness after World War II that the U.S. federal government and many state governments paid attention to the problems of waste management at all. As this analysis of New Jersey's waste history shows, the result was that new sets of regulations, financial models, and collection and disposal technologies were installed on top of what was already in place. This pattern would be repeated countless times as new technologies for disposal emerged (for example recycling, or waste-to-energy incineration) and more precise categories of waste materials were articulated (for example yard waste, construction and demolition debris, food waste). Even the most sincerely crafted comprehensive waste management plans—including many of those generated in New Jersey by state agencies, county authorities, and town governments—could <u>never</u> come to be fully and completely implemented against the backdrop of shifting political sentiment and technological change (to say nothing about diminishing municipal finances).

Nevertheless, as the twentieth century progressed an unusual new obstacle limiting progress toward improving waste management infrastructure emerged: adequacy. At certain points in time it was very clear that New Jersey faced a literal crisis in waste disposal: for example when the day came that all remaining town dumps would be closed, there was understandable panic about where exactly the garbage would go. Yet these moments of crisis featured focused minds and strong decisions. In contrast, during the period of relative calm that has characterized waste management in New Jersey in the first part of the twenty-first century, a collective attitude has crystallized that waste management systems in the state are "good enough." We have an entirely adequate system of waste management infrastructure in New Jersey. Yet most in the world of waste management here would agree that aspects of the system need improving, from regulations and financing to technology and public outreach efforts. The *problem* of garbage is not so bad as to command the type of focused attention and public action that it did in years past, yet we all know we can do better.

The third and final argument in this book offers a path forward, while also bringing us full circle. New Jersey's systems of waste management need a swift kick in the recycling bin: we ought to embrace the twin historical realities of being both a regional epicenter for trash disposal and an innovator in waste management policy to become *the* global leader in materials recovery, processing, and remanufacturing. The foundations to achieve such a goal are already in place:

1 a competitive marketplace for waste management services, including many businesses, large and small, public and private, that already collect and process waste materials (including materials that other jurisdictions have long ignored);

2 a tradition of realistic, collaborative approaches to public oversight and an observable public commitment to supporting recycling programs;

3 a robust collection of road, rail, and port infrastructures serving the state alongside an impressive industrial heritage;

4 a geographic location at the heart of both national and international commerce that would allow innovative new materials derived from waste to reach markets around the globe.

Additional work is needed to advance this vision. Instead of chasing dying industries, public and private investment ought to focus on new waste processing and manufacturing technologies; instead of killing innovative technologies by a thousand regulatory cuts, adopt a catholic approach to incubating new firms that deal in waste collection, processing, brokering, and financing. A prosperous garbage future for New Jersey would require establishing supports for the markets for waste management goods and services as well as streamlining data collection, permitting, and regulatory processes. Most important, however, would be a shift in mindset. Reorienting our thinking about waste management from a problem to an economic opportunity bridges the intellectual gap between comprehensive management plans that are insensitive to local needs and sensibilities and waste systems that are too fragmented to work well.

In any event, before wading too deep into the history of waste management in New Jersey or speculating too wildly about its future, a vitally important question has to be answered.

How Much Trash Are We Talking about, Anyway?

The U.S. Environmental Protection Agency (EPA) estimates that Americans produce about 4.5 pounds of waste per person, each day, totaling to nearly 260 million tons per year in 2014, the last "data year" for which the EPA made an estimate of national trends.[9] It is very important to note that this figure is an estimate of the types of materials that are collected from homes and businesses. Such municipal solid waste (MSW) excludes the waste materials that come out of factories and other industrial sites, as well as sewage sludge from wastewater treatment plants and debris from construction or demolition sites. While estimates for industrial and other types of waste vary widely and are quite hard to pin down, most everyone agrees that the amount of industrial waste far surpasses the amount of MSW.[10] In any case, the amounts have been increasing steadily since 1960.

In New Jersey, the Department of Environmental Protection (NJDEP) has collected data on waste and recycling trends for decades—information which has played an important role in this project, as discussed below. Like the EPA, NJDEP defines MSW as the materials collected from homes, businesses, and institutions like schools; however, the NJDEP also offers the public estimates of total solid waste—including most materials like sewage sludge and construction debris—before breaking out figures based on whether materials are recycled or not.[11] With that in mind, we can say that New Jersey generated 20.8 million total tons of solid waste in 2014, of which 9.6 million tons was classified as MSW. That amounts to about 5.8 pounds per person, per day—significantly more than the national average. However, NJDEP also reports that about 4 million tons of the MSW were recycled, about 41 percent. Though this is short of the state's 50 percent MSW recycling target, it still far surpasses the national average recycling rate of 34 percent reported by the EPA.[12]

Just for comparison's sake: our neighbors in Canada produced about 28.2 million tons of waste in 2014, or about 4.3 pounds per person each day. The European Union as a whole generated 270.5 million tons of MSW in 2016, averaging 2.9 pounds per person, per day averaged across all member states. But within that average, there is wide variation: for example, in Denmark, each person produced about 4.7 pounds of MSW per day, while in Romania it was just 1.6 pounds. To round out the comparisons, in Japan, a 2015 study estimated a total waste generation of about 49.5 million tons, or about 2.2 pounds per person, per day.[13]

It is important to take statistical information about solid waste with a large grain of salt. Differences in the ways that waste is defined, as mentioned earlier, translate into differences not only in how waste is processed, but also how it is accounted for. The particular methods for collecting waste statistics have to be considered in great detail before we can draw precise conclusions from comparisons across countries, and even across states within the United States. Even thought we might be somewhat skeptical about the precision of waste statistics, it is clear that there are great variations in waste production across places and cultures. This is one reason why scholars and researchers have been interested in studying waste management for many years. How can we explain differences in waste generation and disposal practices, if we can explain them at all?

How Can We Examine New Jersey's Waste Management Infrastructure?

There are probably as many approaches to studying waste management systems as there are definitions of waste itself; as such, studying waste management infrastructures is an interdisciplinary endeavor crossing the boundaries of history, anthropology and the social sciences, and policy and economics research. This project follows the models offered by scholars like Martin Melosi, Samantha MacBride, and Richard C. Porter, and like their work, is built on a few different sources of information.[14]

First and most important to this study are historical materials relating to waste management in New Jersey. These include government and public agency documents, news media, and the materials of private firms and citizen groups engaged with waste management issues in the state. The majority of sources were produced between 1960 and the present. Government documents include sources like reports, memos, meeting minutes, press releases, correspondence, requests for proposals (RFPs) and other artifacts of the procurement process. An important consideration in this analysis of government materials is the way in which different components of government have interacted with one another, especially because waste management policy making in New Jersey has been marked by efforts to square visions of a high-technology, ecologically benign waste future with the limitations of county and municipal governments.[15] Public entities engage the waste stream both directly, through collection and disposal work, and indirectly, by setting the rules for collection and disposal. If nothing else, government (at some level) is responsible for financing waste management services (to some extent), and for that reason alone exerts considerable influence over waste management infrastructure and operations. This book presents findings from historical research into governments' roles in waste management drawn from sites around New Jersey, ranging from state and agency archives and libraries (in particular, those of the NJDEP), to county and local government offices tasked with planning and implementing waste management practices. Additionally, the holdings at the Rutgers University libraries, and especially the Sinclair New Jersey Collection, as well as the holdings at the library of the former Meadowlands Commission, were explored and utilized in this aspect of the project. Specific county and local government sites were examined to represent *typical* New Jersey communities, and include a suburban county in North Jersey (Morris), a rural county in North Jersey (Sussex); a rural, coastal, tourism-dependent county (Atlantic); a large county in South Jersey spanning urban, suburban, and rural areas (Camden); and a special multimunicipality unit of government outside of New York City, in an ecologically sensitive area called the Meadowlands (for which the main oversight body is now called the New Jersey Sports and Exposition Authority, or the NJSEA).

News media sources were valuable as a register of public perspectives on waste in New Jersey as well. Counties and towns have historically had their own news media source(s); complementing these are media reporting on issues affecting the state as a whole. New Jersey-focused news media offer considerable space to both environmental issues and the conveyance of readers' perspectives via letters to the editor and user comments, making news media a valuable source of different types of perspectives on waste issues. Both local and state-wide media were examined for this project. Two smaller sets of documents were also examined: those of private companies and those of environmental interest groups. Private companies have long been a significant force shaping waste management policy and practice in New Jersey.[16] Private companies impacted waste management as both landowners, contractors, and facility operators. Private firms play a crucial role in waste management not only as waste haulers and disposers but also frequently as the actors producing state and county planning documents as well as running many landfills, recycling centers, and incinerators. As such, private companies have played an important agenda-setting role in New Jersey. Examples of private firm sources used in this project include shareholder reports, solicitations, other artifacts of the public procurement process (including documents prepared for governments), and testimonies in legal hearings and state and county government meetings. Environmental interest groups in New Jersey have been important for their role in agitating for more ecologically progressive approaches to waste disposal (amid other environmental concerns), frequently acting as a "moral compass" on pollution prevention issues. Thus, their publications frequently offer a valuable critique of government and private entities involved in waste management planning and facility operations.

Readers will observe my preference for letting the historical record speak for itself, through direct and sometimes lengthy quotation of original sources. I do my best to provide context for the words, ideas, and observations of these folks who lived the historical moments in question. It is my hope that this will in turn allow readers to better understand the decisions and perspectives in their own historical moments.

Along with historical documents, the second major source of information for this project is a set of interviews conducted with figures in the history of waste management in New Jersey. Seventeen different semistructured interview sessions were held with twenty different individuals, each having different roles in the past and present of garbage in New Jersey. Each interview lasted between forty-five minutes and two hours, resulting in more than 450 pages of transcribed text. Participants included former chiefs of staff at the NJDEP, directors of waste management departments at NJDEP, county-level utilities authority executives and officials within local government, current and former elected officials, and executives and representatives of private companies operating collection and disposal services in the state. The insights from these interviews were highly informative, and truly fascinating—in the current age of outrage and savage critique toward nearly everything, the interviews are an important reminder that environmental infrastructures are, still, primarily a human venture, and cannot be understood simply as the result of abstract engineering formulas and financial modeling. For the vast majority of the time, these systems are designed and operated by people trying to do their best. The seemingly sterile world of environmental policy-making and infrastructure planning is actually imagined, implemented, and operated by human beings with unique perspectives on the work they do and why it is important for society at large. By and large, the interviews helped shape the overall narrative of the book, and it is only on a rare occasion where individual interviews are quoted and cited in the text.

The third and final source of information for this project has been quantitative data about waste itself. For decades, NJDEP has been collecting so-called scale data for all the waste disposed in New Jersey. That is to say, every time a hauler truck rolls into a landfill, incinerator, or other disposal facility, that truck has been weighed for the purposes of assessing the correct price-per-ton fees. The tonnage data collected by disposal facilities was in turn passed along to the state of New Jersey, which would use the data for analysis and planning purposes. Similar information about recyclable materials was also collected as a key decision-making tool in a grant program aimed at supporting recycling efforts in New Jersey's towns. As a result, there is a very interesting, if at times incomplete, record of the amounts and types of waste generated and disposed in New Jersey. This body of quantitative data, comprising over a million individual records and stretching back to approximately 1993 (as far back in time as we were able to collect for this project), is not the star of this book, and I make a conscious effort to minimize the dizzying array of statistics and quantities that frequently appear in research about waste management. As such, readers interested in learning more about this pool of data can access a scholarly journal article about it (published with the help of some excellent Rowan University students), "New Jersey's Solid Waste and Recycling Tonnage Data: Retrospect and Prospect" published in the journal *Heliyon*, available from https://doi.org/10.1016/j.heliyon.2019.e02313 and also my website, www.jordanphowell.org.

Organization of the Text

The chapters in the book are arranged generally in chronological order, though some exceptions are made to explore particular topics and technologies in greater detail than a purely chronological organization would allow for. Chapter 2 traces the origins of state and county-level waste management planning efforts, including early efforts to investigate poor environmental and economic performance. Chapter 3 examines new legislation mandating a comprehensive approach to planning in New Jersey that emerged in the 1970s, along with the state's unsuccessful attempt to ban disposal of garbage from New York and Philadelphia. Chapter 3 also looks at efforts by state planners to implement a system of directing flows of waste to particular disposal sites, as well as the role of organized crime in New Jersey's waste management infrastructure. Chapter 4 focuses on New Jersey's attempts in the 1980s to move away from landfilling and toward alternative technologies, and in particular waste-to-energy incineration and intensive recycling. Chapter 5 examines the unraveling of New Jersey's waste management system in the 1990s, first by a change in strategy to emphasize recycling above all other disposal options, and then by a series of devastating court rulings. Chapter 6 considers the prospects for the future of waste management in New Jersey. A complicated set of issues faces everyone involved in the waste management community, ranging from the instability and unpredictability of markets for recyclable materials and other waste-derived products, to dwindling public finances; from advances in disposal technologies linked to high-tech startups, to large national and multinational corporations displacing mom-and-pop firms. More fundamentally, the waste management community must navigate the reality that absent a highly visible crisis in waste disposal—something unlikely to occur in New Jersey anytime soon—neither the general public nor policymakers will demand changes to further improve New Jersey's system of waste management. Chapter 6 offers a route forward, making a case for capitalizing on New Jersey's experience with trash while suggesting new policy and infrastructure approaches to waste management that could transform the state into a global leader in innovative collection, processing, and remanufacturing. Running counter to many "zero-waste" pronouncements about the possibility of eliminating waste entirely, this book concludes by acknowledging the reality that as humans, we will *always* be producing waste and requiring new strategies for handling it.

2

Origins of Waste Management Planning in New Jersey

● ●

> I don't have the answers but this problem
> becomes more and more complex. Waste
> disposal—we take for granted it is going
> to disappear somewhere some place. It is
> awful expensive. I see it becoming more
> expensive in the future and there are
> many, many problems that go along with
> this entire situation once you determine
> what method you are going to move in.
> —Hon. Carmen J. Armenti, Mayor,
> City of Trenton, May 16, 1969

New Jersey was originally an agricultural state, dotted by small farms serving the growing cities of New York and Philadelphia. But with the industrialization boom beginning in the nineteenth century, thousands flocked to New Jersey's own emerging cities, former farmers and new immigrant arrivals alike. As these groups established themselves in urban New Jersey they created new cultures, patterns in settlement and economy, and of course, new issues in waste management. Early waste disposal methods were rudimentary—find a patch of seemingly worthless land and toss your waste into it. Alternatively, some waste, particularly food wastes, might be collected by pig farmers and fed to the herd. As New Jersey urbanized, particularly northern New Jersey, some basic types of incinerators were installed, mostly because other open lands were being utilized for housing and industry. Given its position wedged between two large urban centers, New Jersey received considerable volumes of Philadelphia's and New York City's waste as well, none of which was disposed of properly in any environmental sense. In the wake of WWII, New Jersey set down a path of becoming one of the United States' most modern and wealthy states, characterized by suburban settlement patterns and mass-consumer lifestyles. One aspect of this rising wealth was an increase in wastes, particularly plastics, myriad new goods imported from all over the world, and "miracle" synthetic materials made from chemicals. While retaining the moniker of the "Garden State," by the 1960s it was clear that New Jersey's economic future laid not in agriculture but instead in heavy industry, value-added manufacturing, and the increasingly lucrative service sector encompassing the full range of business services from insurance and finance to real estate and endless varieties of litigation.

These new patterns of life and material consumption introduced challenges for government, too, as local and state officials started to worry not only about the environmental impacts of waste disposal but also the finances of managing a seemingly endless and rapidly growing waste stream. Accordingly, the 1960s and early 1970s was when state and federal entities began to involve themselves in what had historically been a local issue, through new regulatory and planning efforts. An anecdote about the state of affairs for the waste industry in the 1960s illustrates why this was so vital. According to George Tyler, a former attorney for the New Jersey Department of Environmental Protection (NJDEP) who participated in an interview for this project,

> [in the late 1960s] we had . . . three or four hundred garbage dumps in this state. Some of whom were municipal and didn't charge anything. Some of which were in the Meadowlands charging $3 a ton. It was an all-cash business. I don't know this for a fact, but the heavy rumor was that it was all mobbed up. And I was at facilities where I watched truck after truck stop at the gate, hand a big roll of bills to a guy at the gate, and then go on in and dump. And who knew where the waste was from, what was in the truck, or how much money had been paid . . .
>
> It was a free for all in the world. You could dump anything anywhere. There were "teepee" incinerators in Newark. Big metal teepees . . . maybe 20 yards across at the bottom. They went up to the top. No air pollution controls. People drive in, push—they plow the stuff in and just keep burning. They burned 24 hours a day.[1]

Even as New Jersey grew in wealth and residents' consumption habits increased, in the 1950s and 1960s there also emerged a new consciousness about the impacts of growth, industry, and waste on the physical environment, inspired in part by Rachel Carson and her book *Silent Spring*. Illustrations of the environmental impacts of these decades of unchecked growth coupled with largely unsupervised waste disposal were unfortunately common in the Meadowlands area, as the swamps near New York City had been used for all manner of disposal for decades. Former New Jersey Sports and Exposition Authority (NJSEA) director of solid waste Tom Marturano explained in an interview that the Meadowlands in particular were

> The Wild West . . . We recently took over a 100-acre site in Kearny that have been filled up in the '50s and '60s, and we . . . did some digging. We found records of pure chromium waste being dumped in there 40 barrels a day for years. It wasn't just [municipal] waste was being dumped. It was industrial waste. A lot of [metal] platers around here. It was all kinds stuff being dumped . . . The trucks would literally come, they would make a depression. They just dumped their oil into it.
>
> [One time,] we literally dug down—we ended up having to condemn a small piece of the land, and we had to show them what condition it was. And we all met out at the site. We had the condemnation commissioners, and we had the property owner, and his attorney, and his engineer. And I said, "Pick three spots. Here's some flags. Pick three spots, put the flag in the ground," I had a big excavator. He did it. Excavator started digging a hole. Well, it was like the Beverly Hillbillies. It was pure oil. We got down about eight feet, pure oil. I mean, you could have refined it.[2]

This chapter examines some of the motivating factors behind why New Jersey has ended up with a largely state-directed, county- and municipality-implemented waste management system whose many actors include a blend of public and private service providers. Moving from the system George Tyler described as a "free for all" to something with controls for environmental pollution, social equity, and reduced potential for corruption in the government bidding process would clearly take time and likely shine sunlight on more than a few skeletons stuffed into closets around the Garden State.

This chapter next examines some of the outputs of this initial era of greater organization and oversight of New Jersey's waste management system, and in particular early state and county efforts at comprehensive planning. These documents reveal the severity of the state's "waste crisis" as well as early accountings of the quantities, costs, and various outcomes of New Jersey's waste volumes. As such they provide invaluable context for the decisions and proposals that were floated in the 1970s and later, and in particular, the notion that a comprehensive and unitary system for managing all of New Jersey's garbage could be devised and implemented.

The Origins of Waste Management Planning in New Jersey

On October 22, 1965, Senator C. Robert Sarcone announced the start of a public hearing of the Joint Commission to Study the Problem of Solid Waste Disposal by reading a segment of the resolution enabling the commission's work: "It shall be the duty of said commission to study the problem of solid waste disposal, and, in connection therewith, to assess the availability to the respective counties and municipalities of this State of land sites suitable to meet the need for disposing of solid waste by various methods, to undertake research into methods of improving procedures and techniques for solid waste disposal and to encourage counties and municipalities to work cooperatively on a regional basis to resolve their common problems of solid waste disposal."[3] Before a panel of state senators and assemblymen from the northern New Jersey counties of Hudson, Bergen, Essex, and Middlesex, Dr. Roscoe P. Kandle, the state commissioner of health, offered an overview of the problems with waste disposal confronting New Jerseyans. The list was long and complex. Over 21,000 tons of municipal and industrial wastes, on average, needed to find a home in New Jersey every day, yet in 1965, 303 of 567 towns did "not have organized refuse disposal areas." Just eleven of the thirty-six incinerators known to exist in the state were operational; the majority of these were "of inadequate capacity to handle the refuse produced by the municipalities they serve." Dr. Kandle noted that the majority of the daily estimated volume was generated by the five counties of northeastern New Jersey, the major disposal sites for which were located in the ecologically fragile Hackensack Meadowlands. At the time of Dr. Kandle's testimony, fires in the Meadowlands landfills were not uncommon, owing to a "large and unmeasured" quantity of "bulky, flammable wastes" from construction projects coupled with severe drought that had turned some landfills to tinder. Organized waste collection was far from comprehensive, and sometimes nonexistent, with contracts for collection and transportation frequently awarded under murky circumstances to bidders of questionable qualifications. A significant component of the daily volume was imported from New York City and Philadelphia. Not unrelated to these collection issues was "indiscriminate and promiscuous dumping of refuse along highways, streets, and nearby access roads to the refuse disposal areas."[4]

In short, the situation for waste management in New Jersey was a mess. However, it was not as though no rules for waste management existed: dumps, piggeries, and incinerators had been within the purview of the State Department of Health for decades prior to formation of the Commission to Study the

Problem of Solid Waste Disposal. In 1957 the State Department of Health had even promulgated rules for waste disposal, publishing Chapter VIII of the State Sanitary Code with exactly two regulations. First, that "disposal of all organic and/or combustible matter on lands in this State shall be made only through use of . . . Sanitary landfills . . . OR Incinerators"; and second, that these rules did not apply to "family garbage or family refuse on the premises where the family resides," nor should the rules be interpreted as permitting disposal of domestic sewage.[5]

For officials like Dr. Kandle the problems surrounding waste disposal originated not in violations of the Sanitary Code, but rather the failure of towns, counties, or the state as a whole to organize and coordinate waste collection and disposal. Describing the attitude of some New Jersey local officials toward waste disposal as "almost a game now [where] municipalities outlaw dumping of solid waste and then they just . . . find some other sucker who will accept it," Dr. Kandle urged the lawmakers hearing his testimony to move forward quickly with implementation of the federal Solid Waste Disposal Act and begin the process of collaborative, long-term waste management planning.[6]

The federal Solid Waste Disposal Act was passed in 1965 with the intention of improving the state of waste management in the United States, particularly by (financially) supporting demonstrations of new collection, processing, and disposal technologies and encouraging states to take a comprehensive look at how wastes were handled within their own borders—in other words, to plan for waste management. These aspects of the legislation were contextualized within broader goals of reducing pollution of land and water resources, and curbing some of the harmful health impacts of town dumps. New Jersey officials acted on Dr. Kandle's advice, and were able to secure a grant from the federal government that same year, to undertake a comprehensive study of solid waste in the state. The result was *The New Jersey Solid Waste Disposal Program*, released in 1968 by unnamed authors within the State Department of Health.[7] This document contains significant concepts—and identifies major problems— that would shape New Jersey waste management for years to come.

The *Program* declared that "neither effective solid waste management nor meaningful management planning is feasible for individual municipalities," but simultaneously that "New Jersey is one of relatively few states . . . which is wholly municipally incorporated . . . the county in New Jersey has responsibility for only a limited range of functions, and substantially no supervisory authority relative to its component municipalities."[8] Given the history of cities and towns having to manage wastes on their own, any attempt at county- or state-level waste management planning started at an immediate disadvantage: "to attempt to formulate a state-wide optimum solution for solid waste management—for solid waste disposal sites and facilities—and attempt to impose that plan upon each of the 567 municipalities; upon each solid waste collector, public and private and upon each disposal site operator, public and private . . . more than two thousand individuals and corporate entities . . . was obviously impossible."[9] Accordingly, authors of the *Program* sought to develop strategies that would "encourage cooperative solid waste management planning and subsequent solid waste management by groups of communities—several contiguous municipalities acting cooperatively through their county government; or municipalities operating in multi-county groups."[10] However, cooperative municipal action— towns agreeing to help each other directly—"was generally viewed with skepticism by many municipalities," leaving *Program* authors to conclude that the clearest path forward for organized waste management planning in New Jersey would be for counties to take the lead role.[11] With this type of structure, *Program* authors reasoned, the state government could also propose birds-eye-view solutions for the whole state, that could be handed down to county governments sharing implementation responsibilities with towns.

Underlying the *Program*'s approach to waste management planning were commitments to extensive quantitative data gathering and analysis as well as coordinating waste management with other comprehensive planning exercises in the state and region. According to the authors, waste management planning in New Jersey "must rest solidly on a base of data—data about the people of the state, its industry and commerce, and its land . . . analysis of . . . type of refuse . . . generation rates and their change over time . . . suitability and capacity of existing disposal sites . . . the projections and forecasts required to develop short- and long-range solid waste management plans . . . insure continuing program responsiveness to changing or unanticipated conditions or innovations."[12] Such data would be gleaned from U.S. Public Health Service surveys and other public sources, and also from original State Department of Health studies focused on determining baseline conditions of waste volumes and types, and inventories of disposal facilities around New Jersey. Fearing that the *Program*'s efforts could be dismissed if they came into conflict with plans for solving other problems, the authors also encouraged close coordination of waste management planning activities with other emerging state-level planning concerns like water and air pollution control, recreational programs, and ongoing transportation and housing studies.[13]

By highlighting a potential leading role for counties, the importance of quantitative data in waste management planning, and the coordination of waste management with other planning activities, *The New Jersey Solid Waste Program* introduced concepts into New Jersey's approach to waste management that would be consequential for decades to come. At the same time, the *Program* left a considerable set of questions unexamined. Most immediately, the *Program* focused attention exclusively on northeastern New Jersey and the dense, urban counties abutting New York City. This is understandable in some ways, given the worries about the impending development of the Hackensack Meadowlands negatively impacting capacity for waste disposal in that region. Where would the garbage go if not to the dumps and landfills of the Meadowlands? However, devising a statewide plan for waste management that subjugated the western, central, and southern parts of the state to second-tier status might not be politically palatable. One potential workaround would have been to elevate each county's government to a key role in studying and planning for waste management, thus allowing the state agencies to take a broader view while allowing more local units of government the opportunity to develop strategies unique to their own situations. The *Program* noted however, at the time of writing, that there were no legal or governance mechanisms in place by which such a county-led approach could be implemented.

The approaches to more effective waste management planning identified in the *Program*, as well as the barriers to their implementation, were examined in greater detail during a series of hearings held in 1969 by the Special Legislative Commission to Investigate Certain Problems Relating to Solid Waste Disposal, a joint New Jersey State Senate and Assembly panel. Considering at some length the testimony offered during the hearings is instructive, and reveals the fascinating textures of municipal waste management operations in New Jersey at the end of the 1960s. The polarity of opinion regarding the optimal role for local government in handling garbage is also revealed. These hearings included insights and commentary from the mayors, attorneys, and administrators of cities like Trenton, the state capital, and Kearny, a Meadowlands town with numerous landfills, as well as executives from the New Jersey League of Municipalities, an association and advocacy group for local officials. Many of the hearing participants admitted to knowing little about the actual quantities and composition of their town's waste stream, and even less about how prices for waste collection and disposal should be set by contractors. A town commissioner from the Township of North Bergen, Charles J. Weaver, suggested that the previous efforts by the State Department of Health to limit dumping and smoke-belching incinerators, while effective in reducing pollution, had also limited the number of disposal options for towns and thus inadvertently introduced waste management monopolies. This made impacts both on the costs of disposal and collection service, as Commissioner Weaver explained:

> The disposal of waste matters is only a portion of the garbage problem . . .
> collection practices are also of great concern. Many municipalities have
> resorted to private contracts over the years. While the cost for such contracts
> has risen due to inflation . . . [some] contracts have risen because existing
> dumping sites have been closed either due to health department regulations or
> lack of adequate land. This lack of adequate landfill sites has narrowed
> competition in the garbage contracting field because a prerequisite in the bid
> procedure is proof of an adequate disposal area for the contractor's use. Thus,

many of our municipalities in the metropolitan area are receiving only single bids when they advertise.[14]

Commissioner Weaver, also testifying on behalf of the New Jersey League of Municipalities, called for waste collection and disposal to be treated as a regulated utility, with prices and profit rates controlled by the State Public Utility Commission. Furthermore, Commissioner Weaver suggested that before contractors submit bids to provide waste service to a town they be somehow licensed or qualified by the state. Both ideas were seemingly well received by the lawmakers participating at the hearing as being sensible approaches to control municipal costs.

The rationale behind Commissioner Weaver's suggestions would be illustrated by the testimony presented by Joseph M. Healey, the mayor of Kearny, New Jersey and Paul McCurrie, the town attorney. Located in the rectangle outlined roughly by Newark, Jersey City, Fort Lee, and Passaic, and very near New York City itself, the marshy landscapes of the Hackensack Meadowlands had long been attractive disposal solutions for the businesses and residents of this densely populated area. Kearny was no exception. Mayor Healey relayed that while Kearny had built an incinerator in the years between the First and Second World Wars, the performance of the facility was so poor that it operated only for a few months before being mothballed. The mayor described how marshlands in Kearny had been used as dumping grounds for Meadowlands-area residents and industrial operations alike, including shipbuilding and other military-affiliated activities during World War II. In order to try and manage the flow of wastes, the town established four designated dumping sites. The mayor explained that for businesses in Kearny, disposal at the town sites was free, and residents had to pay only a $2 fee for a permit for unlimited disposal. The four sites were leased, starting in 1949, to contractors at the rate of just one dollar per year, and the contractors were free to charge whatever prices they liked to those coming from outside Kearny to dispose their wastes at the site. The town reserved the right to sell any of the land, at any time, and when the land was filled in with waste to a particular height, it would automatically become town property again and salable as an industrial site.[15]

While the meadowlands area had long been considered worthless swampland, by the 1960s the area was attracting attention from environmentalists, and in 1969 contemporaneous with the hearings on waste management, the New Jersey Meadowlands Commission was established to protect the area's natural resources and limit future development of the marshes.[16] Prior to the new commission taking any action, Mayor Healey explained, Kearny officials decided to put out for bid new contracts to operate the dump sites. They had included in the bid qualifications that bidders demonstrate at least five years' worth of experience operating a waste dump (and also that bidders include their criminal records, and "furnish details thereof, including nature of crime, date and place of conviction, and name of court and punishment resulting"[17]). Five bids were ultimately submitted and the winning contractor, a joint venture called the Municipal Sanitary Landfill Authority, offered Kearny a jaw-dropping $516,000 for a ten-year lease, with the town retaining all the previous rights to sell and convert the land to industrial use. The new contract represented an increase in value of more than 50,000% per year to the town—which would transition from receiving $1 per year from contractors, to about $142 *per day* over the life of the lease.

In the mayor's words, "provisions in it [the lease] are obviously very favorable to the landlord [the town of Kearny]. It's a good lease."[18] While proud of the arrangement he and Kearny attorneys had made because of the income it would provide the town, when pressed by the legislators conducting the hearing on the specifics of the deal and the qualifications of the winning bidder, an interesting set of circumstances emerged. Essex County assemblyman Kenneth T. Wilson probed Mayor Healey about the contract's pricing provisions:

ASSEMBLYMAN WILSON: What did you feel would be a good bid for a 10-year period? You must have had some idea?

MR. HEALEY: I don't think anybody had any idea, to be perfectly honest with you. We were going from a one-dollar bid to something - we didn't know.

ASSEMBLYMAN WILSON: You didn't have any idea of what would be a good price?

MR. HEALEY: No.

ASSEMBLYMAN WILSON: Would you have accepted anything; for example, if the low bidder was $100,000, would you have accepted that?

MR. HEALEY: That would have had to be referred to Council and then there would be a discussion on it. But the Council felt that the bid as presented, the highest bid, was a good bid. I'm sure if they didn't feel that way they wouldn't have voted for it.

ASSEMBLYMAN WILSON: Well, you had no established figure before?

MR. HEALEY: No, sir . . .

ASSEMBLYMAN WILSON: . . . What amazes me, for example, is that the high bidder was $216,000 over the next bidder. To me that's a tremendous difference. It seems that the other bidders were all relatively close but then the top man was so far ahead, over $216,000, and I was just wondering if there was a price that you had actually established that you felt would have been fair to the municipality, but you have indicated that there wasn't any.[19]

Given the provisions of the deal, and in particular that Kearny residents, businesses, and the town public works department could continue to utilize the site for free, but that the contractors would be able to charge others any price they desired, it is perhaps not surprising that the profit margins for contractors or bidders were completely opaque. As Senator Milton A. Waldor discovered,

> SENATOR WALDOR: So you have no idea nor has the municipality as to whether or not (1) it is a profit-making endeavor or not and (2) if it is a profit-making endeavor, what the amount or percentage of profit might be by the operators of that dump site. You don't know that?
>
> MR. HEALEY: No, sir . . . I couldn't conceive that a group of people could by spending an hour a day or even a day determine how much money would be paid into the dump area . . . I don't know how anybody could answer that except the people who had the lease.[20]

Given the apparently lucrative market value of operating a dump site, legislators conducting the hearing questioned the mayor and the town attorney as to why Kearny did not simply operate its own disposal sites and keep all the profits, rather than settling for the fraction that would be paid by contractors. In response,

> MR. HEALEY: . . . there has been talk about a municipality like ours going into the garbage disposal business. To me, I feel that this would be a very difficult thing. I have made some studies of garbage disposal and for us, as a community, to start a sanitary landfill and have to meet the requirements set forth at this particular time under the State Department of Health, and for us to have to hire engineers through Civil Service and pay the rates to our laborers - whereas these people [the contractors who had submitted bids] seem to be able to operate dumps - the men who work on the dumps actually are not I would say, the highest paid. It would be difficult for us more so to keep the landfill under the - . . . for us to go out and look at the fill would require a whole new department. And I felt, in view of the fact that this land in the near future would become part of the Meadowland Development Agency [sic] and that it would cost, it is estimated by our people, some two hundred thousand dollars for the Town of Kearny to go into this business - I felt that it was for our best interest to bid them, not lease them for a dollar a year.[21]

In particular, he related the experience of North Arlington, a town nearby Kearny and also in the Hackensack Meadowlands, which was apparently struggling to meet obligations relating to its town dump and attempting to explore a transfer of the facility to Bergen County government:

> MR. HEALEY: I have a copy of the *North Arlington Leader* . . . you asked if we had done any investigating . . . It's in regard to the Bergen County Freeholders who have had some experience in four or five dumping sites, I

believe, and they tell about how the price of dumping has increased and how it's almost impossible to operate as a Board of Freeholders; in fact, it says here in the *North Arlington Leader* of January 23rd that they must maintain equipment of over $200,000, and tells about the cost of bulldozers and what not. They say that one alternate studied by the Freeholders is to put the dump operation under private control. They have already discussed the possibility of using a company to operate the new operation to be opened at Lyndhurst. This is one of the things that interested me because we had done some checking on this and taking in the cost and what not, we felt it was not feasible for us. There have been many, many articles written on this situation. In the North Arlington papers there have been a few editorials.[22]

The legislators presiding over the hearing seemed sympathetic to the notion that the heavy, somewhat unpredictable costs of designing and operating a proper sanitary landfill could motivate a town to contract out something like waste disposal services. But questions lingered over the nature of Kearny's selection process. Lawmakers present wondered if towns elsewhere in New Jersey would have been eligible as bidders, in effect leasing land in Kearny for their own waste disposal purposes, while also generating revenue by charging still other towns and businesses to use the site. Senator J. Edward Crabiel, Assemblyman Frank J. Dodd, and Senator Waldor all questioned the mayor about implications of such a scenario:

SENATOR CRABIEL: What would your reaction be, Mayor, if the Legislature passed legislation which would allow municipalities the right of eminent domain?

MR. HEALEY: In other communities? I think that's a legal question.

SENATOR CRABIEL: [...] there are some municipalities that have no dumping facilities. Several people testified before us previously that they were boxed in because they had no place where they could dump in their own municipality, and I think it is agreed that in some municipalities, that would be impossible. The alternative would be to give them permission to run a dump under the control of the Department of Health in some other municipality. I wondered what your reaction would be to that, not only as being a former president [of the New Jersey League of Municipalities] but the Mayor of a community.

MR. HEALEY: I think most Mayors believe in home rule. I don't feel that a municipality has a right to go into another municipality.[23]

Ostensibly to avoid attracting inexperienced contractors, the mayor explained how the town's call for proposals required bidders to demonstrate their waste management qualifications by submitting documentation of at least five years' experience operating a landfill. At the same time, the call for proposals did not specifically bar a municipality with the necessary experience from bidding on the Kearny sites:

SENATOR CRABIEL: Did you have a reason, Mayor, for such specifications so as not to allow a municipality to bid?
MR. HEALEY: Our specifications were not set to eliminate municipalities.
SENATOR CRABIEL: I think you stated that they had to have five years' experience, so this would stop the municipality that presently isn't running a dump. I guess perhaps under your specifications, it would have allowed a municipality who isn't running a dump to bid.
MR. HEALEY: I'm sure we would have, yes.[24]

Mayor Healey's position was that a town which had been operating a dump site directly, for at least five years, would have been eligible to bid on operating the sites in Kearny. However, as Mayor Healey himself had explained, few towns were willing or able to take on the risk and expense of operating a dump, and those that did would likely not need access to a disposal site anyway. Senator Waldor and Assemblyman Dodd, however, continued with the line of questioning:

SENATOR WALDOR: Incidentally, Mr. Mayor, do you know - I don't whether you do or not - who the principals of the Municipal Sanitary Landfill Authority [the successful bidder for the Kearny site] are . . .
MR. HEALEY: You mean, the Municipal Sanitary Landfill?
SENATOR WALDOR: Yes, sir, the people who were the successful bidders for that land.
MR. HEALEY: William Keegan, Incorporated.
SENATOR WALDOR: William Keegan, Inc.?
MR. HEALEY: Yes . . . 411 Bergen Avenue, Kearny. I am reading this from the questionnaire. Officers - William A. Keegan, Jr. and William A. Keegan, Sr. Reclamation and Improvement Company, 90 Milburn Avenue, Maplewood, N.J. - officers Michael Cignarella, Salvatore Nesto. Peter Roselle & Sons, 163 Tremont Avenue, East Orange, N.J. - officers Louis P. Roselle, Crescent J. Roselle. Delaware Sanitation Company, 280 Central Ave., Orange, N.J. - officers Joseph Cassini, Jr. and Anne Cassini. Those are the four.
SENATOR WALDOR: Those are the four corporations that comprise the Municipal Sanitary Landfill Authority to whom the bid was awarded.
MR. HEALEY: Yes.[25]
[. . .]

SENATOR WALDOR: Again, part of the specifications that were advertised, I take it, as Senator Crabiel indicated through his question, was that any bidder, in order to qualify, must have had five years' experience in the operation of a dump site. Is that correct, Mr. Mayor?

MR. HEALEY: That is correct.

SENATOR WALDOR: And, therefore, would that not, in your judgment, disqualify any municipality, for example, who had never been engaged in municipal collection or in any form of collection or operation of a dump site, from bidding under the terms of these specifications? They couldn't qualify if they hadn't been engaged in the operation -

MR. HEALY: That is correct; it would disqualify them if they had not had that prior experience. So the Town thought it was important to put that in, in order that the sanitary landfill program was conducted in the proper manner. I understand that it takes a great deal of experience to do this properly.

SENATOR WALDOR: I'm not questioning the reason for putting it in. I just want for the record all of this straightened out. Just let me ask you this: The people who were the successful bidders this time, and you named them to us, were they the same individuals who had operated the dump site under the dollar a year lease prior to that, for the previous 20 years?

MR. HEALEY: Yes.

SENATOR WALDOR: And was it under the same name, the Municipal Sanitary Landfill Authority?

MR. HEALEY: No, sir, they were under four individual names.[26]

[. . .]

ASSEMBLYMAN DODD: Under the five-year experience provision of the dump operator, would it be possible for a town like Orange to sub-contract the dump operator in order to qualify—

MR. MCCURRIE: I would think that would have been proper.

ASSEMBLYMAN DODD: Or the Town of West Orange, or any town for that matter, could qualify by sub-contracting a dump operator . . . who maintains its own sanitation—

MR. MCCURRIE: [interrupting] Yes.[27]

In effect, Kearny had awarded the contract for operating their dump site to a partnership also serving as, or aiming to serve as, waste disposal contractors for other towns in northern New Jersey. While Mayor Healey had in his testimony initially rejected the possibility of other towns directly owning dump sites in Kearny, through a network of contracting arrangements, the town government had knowingly or unknowingly permitted exactly that to happen. And, since neither Kearny nor the state of New Jersey controlled the rates charged by a landfill operator to dispose of waste at a given site, these landfill operators and client towns would have essentially unlimited landfill access of their own while being able to charge others any price they desired for disposal service, while also avoiding the obligation to build and operate disposal sites within their own town borders.

The episode surrounding the dump site contracts in Kearny is complex and somewhat opaque even after examining the historical record. But it is illustrative of many of the challenges and approaches to contracting arrangements that were shaping waste management policy and practice in New Jersey at the end of the 1960s. On one hand, town officials like Mayor Healey were strongly opposed to other municipalities using their disposal sites, a stance which would seemingly limit the possibilities for cooperation in waste management planning. On the other, contractual arrangements that achieved essentially the same outcome were welcomed for the income they generated for municipal budgets and the appearance of control they demonstrated over a growing waste stream.

Other public officials offering testimony at the 1969 hearings argued positions directly counter to the representatives from Kearny. John Dill and S. Thomas Gagliano appeared before the hearing to represent the Monmouth Shore Refuse Disposal Committee, a budding consortium of twenty-one municipalities largely, though not exclusively, in Monmouth County, that had banded together on the recommendation of the Monmouth County Planning Commission to search for and secure a disposal site that could be used by all committee communities. Dill and Gagliano explained that New Jersey law would permit the communities to jointly own and operate a disposal site in one of the member communities, but not to purchase or lease something in a town that was not participating in the committee. Furthermore, the pair pointed out that the member towns would have to be unanimous in their decision about where to locate the disposal site—each town held veto power over whether the collective's landfill could be within its borders or not. Perhaps most unusually, Dill and Gagliano noted, the legislation enabling communities to form an Incinerator Authority might also permit communities to jointly operate a landfill, but only so long as none of the towns participating were themselves *collecting* waste from homes and businesses within their borders— since that could lead to a scenario where member towns were competing against the functions and finances of the authority, even though waste collection and disposal were two very different problems.[28]

Their testimony supports the conclusion that a county-led system could efficiently impose a waste management plan, complete with disposal sites and collection services, over the protests of individual constituent municipalities. Yet, the pair expressed a preference for waste planning to remain the responsibility of towns:

MR. DILL: I think a county line is really of no significance in the engineering problem of collection and disposal of garbage . . . I feel definitely that the collection should be a municipal responsibility and not a regional responsibility or a county or a State responsibility.

SENATOR CRABIEL: I agree with you there.

MR. DILL: In our case, in Monmouth County, there are four areas composed as units in this county study, one right adjacent to the Raritan Bay area. I see no reason why they should arbitrarily limit any joint action to a site within the county. And on the other hand, we are adjacent to Ocean County and I see no reason why Ocean County municipalities should not operate in the same disposal areas as we do.

SENATOR CRABIEL: Except I think we are going to have to crawl before we walk, and it's going to be hard enough to get it within the county, let alone getting it across the county lines.[29]

Ever more contrarian positions emerged as the hearings proceeded. Dr. William F. Westlin, mayor of Chatham Township in southeastern Morris County—a rapidly suburbanizing area of northern New Jersey—explained that the town had recently closed its landfill simply because residents no longer wanted one in city limits, and not because of any particular problem with the operation. Dr. Westlin argued that duties for waste management planning should be shared, simultaneously, by county government and multimunicipality authorities that cross county lines:

DR. WESTLIN: It is quite obvious that nobody loves a dump . . . As to the question of whether authority should be given to the county or to a landfill authority, I would agree that it probably should go to both. We live in an area—we live in Morris County, the southeast section of Morris County, but we have many interests in common with our neighboring communities in Union County. I attended a meeting on Tuesday night of three communities of Morris County and four communities of Union County, and one community of Essex County, and all of us feel that we have a common problem and we could very well form a joint meeting or authority to handle our problems. However, it's also likely that it would be necessary for us to select a site in a county other than the three that I mentioned, and this apparently would not be possible under existing legislation.[30]

Dr. Westlin, like most others offering testimony, asserted that waste collection and disposal were two distinct problems, and that shrinking disposal capacity as towns like his own closed dump sites, whether for political or public health reasons, was also the root of any problems with collection. In tacit support of the approach taken (intentionally or unintentionally) by Mayor Healey and the town of Kearny, Dr. Westlin was in favor of giving multimunicipality landfill authorities the ability to purchase or even condemn land anywhere, for waste disposal sites, and also for towns to choose to close landfills for any reason as the residents of Chatham had done:

SENATOR CRABIEL: You closed the site in your own township but you don't favor your right to condemn a site in some other place and, as I understand it, you aren't particularly satisfied with a county operation doing this ... Now I am trying to find out where you think the dump site should come from.

DR. WESTLIN: I think it may be possible for communities to find dumping sites outside of their own community and purchase these sites and operate them. But I think as long as it's possible for that local community at any time in the future that the community which is selected for the site - let's postulate a hypothetical situation in which there is a sparsely populated area and there is no objection at the present time to a group of communities establishing a landfill site. Now as time goes on, the population of this community increases, the type of home that goes in there is of a higher caliber than the type that is in there at the present time and of higher value, and the citizens of the community begin to agitate with regard to the presence of this landfill in their community. It has happened in our own ...

SENATOR CRABIEL: Suppose another community decides to buy the present site in your town where you just stopped the dump. You wouldn't allow that.

DR. WESTLIN: I'm afraid that it would create tremendous problems.

SENATOR CRABIEL: I know that, but how do you propose then to be able to turn around and buy a dump site in some other community and not allow them the privilege of closing that dump by ordinance as you did. I don't quite understand what you're proposing. You appear to me to want your cake and eat it too. I mean, you want to keep your town clear but you want to buy it somewhere else, but you don't want them to have the same privilege that you have.

DR. WESTLIN: I think that if we allowed a community to condemn property in another community, then every community could then pick on the opposite community. I think that when you establish an Authority, a group of cooperating communities, that they should have the right to condemn property in another community, whether it's among the group of communities that have joined together in the Authority or not.

SENATOR CRABLEL: [...] Do you agree, though, that the community where the dump is going to be condemned, that if that community is a member and it vetoes it, it should be vetoed?

DR. WESTLIN: No.

SENATOR CRABIEL: Then let me ask you this: If your town was a member of the Authority and the majority decided to use the dump that you presently closed, do you want to give up your right to veto it?

DR. WESTLIN: I think it is necessary for communities to cooperate for the common good, and that sometimes this cooperation may work a hardship on one or more of the communities that have banded together. I think this is the price that we have to pay. I think there is a difference between allowing a single municipality to condemn another municipality whereby, as I said, you could have ten communities with 10 landfill sites in every one of the other communities, but it seems to me that when a group of communities combine together in an Authority and this Authority studies the available sites and determines that it is for the common good to have a landfill site in one of those cooperating communities, this community that has been selected must be willing to accept it . . .

SENATOR CRABIEL: Even though that might in effect put the dump back in your township?

DR. WESTLIN: Yes. I started out by saying that nobody loves a dump but I also feel that we must have dumps. Nobody wants it in his town but if it is the best thing, then it is something that has to be.[31]

Dr. Westlin also testified in strong opposition of state-level regulation of waste collection and disposal pricing—"If the private scavengers appear to be milking the public, then . . . the municipality can go in the refuse collection business itself"[32]—and cast doubt on the political will of most county government officials for taking action on waste management. At least in Morris County, Dr. Westlin relayed, "the feeling of the Freeholders is that the county should not bear the financial burden for the problems of a few communities . . ."[33]

The joint New Jersey State Senate and Assembly hearings on waste management were spread across four sessions in 1969. The initial sessions featured testimony from officials and waste industry participants speculating about the potential and pitfalls of regional cooperation for waste management. But the fourth and final session, focused on the experience of Trenton, the state capital, featured Mayor Carmen J. Armenti, along with city business administrator John N. Matzer Jr. and the director of Trenton public works department, Lewis Klockner, relaying the city's unsuccessful efforts to pursue and implement a multimunicipality cooperative agreement.[34] Mayor Armenti described how he had approached various Trenton suburbs to explore a cooperative waste management plan, but that nothing ever resulted because the arrangements each town had were considered good enough, even while all parties agreed that a cooperative arrangement would be preferable in the long run. The Trenton representatives all advocated strongly for state and county leadership in waste management planning, taking the position that towns would never move quickly enough on their own to develop regional plans.

MAYOR ARMENTI: Initially let me say this, if I may: The problem of solid waste disposal not only in New Jersey but throughout the country, I feel, is one of the most complex, one of the most expensive problems facing municipalities today. It is very difficult for many reasons to determine in what direction municipalities should move in the area of solid waste disposal.[35]

. . . I also feel that whenever we have an opportunity to move in any direction in cooperation with other municipalities, we should on a regional approach.

Just recently, the Mercer County Board of Freeholders called a meeting of the 13 municipalities in Mercer County and they have now OK'ed a study of the county for waste disposal. Now this will include all the municipalities to see if we can come up with a regional approach, whether it may be landfill or incineration and I think this is a step in the right direction. Unfortunately, Senator [Waldor], I just feel that time is passing so fast that there is an urgent need for some determinations. I don't have the answers but this problem becomes more and more complex. Waste disposal - we take for granted it is going to disappear somewhere some place. It is awful expensive. I see it becoming more expensive in the future and there are many, many problems that go along with this entire situation once you determine what method you are going to move in . . . [36]

SENATOR WALDOR: Now, Mr. Mayor or Director or Mr. Business Administrator, are there any recommendations by way of legislation referring specifically to the problems that we have discussed that you might have to present to this Commission?

MAYOR ARMENTI: I think what your Commission is doing is of tremendous assistance at this particular time, Senator. I do think it is something that should have happened sometime ago and I congratulate you for taking the initiative in this area . . . [37]

MR. MATZER: You had asked the question about what we could use in the way of assistance. I think one of the things we have followed very closely is the fact that there have been bills before the State Legislature which would provide some State assistance in the area of garbage disposal, solid waste disposal, and in particular some financial assistance for both studies and for construction. . . .

I think possibly at the State level one of the things that could stimulate an area-wide regional approach would be to tie some strings to State aid saying that State aid is available when two, three or four municipalities get together. Such aid could then be available for construction of a series of incinerators or combination of incinerators and landfill approach. But there is a real need for something such as a county study to come up with an estimate of what our needs are and with a combination of approaches.[38]

Initial County and Statewide Plans

As referenced by participants in the 1969 Special Legislative Commission hearings, already a few counties in New Jersey were developing their own draft waste management plans: Monmouth County (1966), Bergen County (1969), Camden County (1970), and Mercer County (1971) along with—at least according to what was discoverable in the historical record—the municipality of Woodbridge (1970).[39] The plans follow a similar pattern in terms of both organization and content: after introductory remarks from a local official or the principal of the planning firm acknowledging the significance and severity of the waste management issue, the plan presents an overview of existing collection and disposal facilities in the county alongside demographics and financial information. The plan projects into the future the amounts of waste expected to be generated and the costs of handling those volumes in the context of several different management strategies, financing options, and disposal technologies. The plan then concludes with recommendations for how best to move ahead to address the solid waste problem, with the caveat that plan conclusions ought to be reevaluated and adjusted in light of changing local conditions.

One reason why many waste management plans followed the same organization and structure was because the 1965 federal Solid Waste Disposal Act supported the development of these documents, so long as they would address certain topics. As a result, a minor, but identifiable, body of literature developed around the idea of helping local officials understand the waste management threat, and articulate and coordinate a response suitable for their jurisdictions. Professional associations for planners, like the American Society of Planning Officials, published "how-to" guides for waste planning and putting together waste planning reports, while engineering societies and research divisions of newly forming environmental agencies like the Environmental Protection Agency (EPA) contributed to manuals for designing, building, and operating disposal facilities like landfills and incinerators.[40]

Examining the 1970 plan for Camden County is illustrative of this early waste management planning literature, both for the specific information offered in the plan but also as a means to identify ways in which ideas from the 1968 *New Jersey Solid Waste Program* and 1969 Special Legislative Commission hearings could be applied in practice. Furthermore, the Camden County plan typifies the waste management problem as it was found in many counties in the late 1960s, offering a practical counterweight to the general remarks that Dr. Kandle had made to the Joint Commission to Study the Problem of Solid Waste Disposal just a few years prior. The Camden County plan opens by noting that "Camden County has not been by-passed by the solid waste crisis," illustrating the severity of the problem with the observation that "refuse production for the next ten years would be enough to cover 400 miles of county roadways to a depth of six feet." Space was running out in county landfills as the amount of waste produced exceeded 6.8 pounds per person, per day. Coupled with rapid (projected) population growth of 3.2 percent per year, "it can be assumed that the problems surrounding the collection and disposal of solid waste will not only increase but there is a good likelihood that such problems will be intensified."[41] There was little discussion about attempting to reduce the amount of waste produced, and so the plan focused exclusively on addressing the problems of collection and disposal. For Camden County, like many counties in New Jersey, waste management problems in the years between World War II and 1970 lay squarely at the crossroads of public financing, land availability, and a fragmented network of municipal governments.

The plan reported that all thirty-five of the municipalities in Camden County had their own systems of waste collection and disposal, with almost 70 percent of the towns having some combination of public and private contract refuse collection. The situation in just this one county of New Jersey was no less complex regarding disposal: "Waste is deposited at one of fourteen different landfills, eleven of which are located within Camden County. Of these fourteen landfills, four are privately leased concerns located within the county and accept solid waste from nineteen municipalities; seven of these are public municipal landfills located within the county and only accept refuse from the municipality in which they are located; three of the fourteen are privately leased landfills located outside the county and accept refuse from nine county municipalities. The three landfill sites used outside the county are located in Gloucester and Burlington Counties."[42] In other words, the plan identified that there was neither cooperation nor coordination among the towns of Camden County regarding waste management. In fact, several landfills accepted waste only from the towns in which they were located. Noting that many of the existing disposal facilities were projected to run out of space before 1980, the plan considered several approaches to future disposal involving some combination of incinerators, sanitary landfills, and a network of "transfer stations," where waste could be loaded and shipped to a disposal site elsewhere. But each of these options was limited by both the availability of acceptable sites and the challenge of financing. For example, the plan notes that the only acceptable location for a large sanitary landfill would be in the southern end of the county in a town called Winslow—the majority of which lies in the Pine Barrens, an expansive natural area that would later become a federally protected nature preserve. The majority of the northern half of the county had already, by the late 1960s, been developed, designated for development, or set aside for some other particular use.[43]

Even if there had been an ideal location to site a facility, the means to finance a county-wide waste management system and especially a disposal facility was unclear. Though in 1970 the projected costs of constructing and operating a landfill, incinerator, or transfer station were within allowable public debt limits for Camden County, this was not necessarily true for the other counties of New Jersey. The plan considered repayment via general tax revenues, special fees, user service charges, or "other financial approaches" including private owner/operator arrangements, where the facility would be guaranteed a certain volume of waste per year and price per ton by the county. While planners could quickly forecast waste volumes and compositions or model traffic routes for collection and disposal vehicles, answering the twin questions of who would pay for waste management, and how, proved far more complicated.

Adding to the uncertainty surrounding repayment was the legal impermissibility of counties in New Jersey, at the time, to develop and operate certain types of infrastructure. In fact, the very first finding of the 1970 Camden County report is that "Legislation which would allow a county and its Board of Chosen Freeholders to operate a solid waste disposal facility does not exist as of this writing."[44] Echoing the testimony offered during the 1969 Special Legislative Commission hearings, nothing in the late 1960s would have allowed county government to legally pursue the type of coordinated development and operation of waste management collection systems and disposal facilities that had been debated in several venues already by this point.

In July of 1970, from the fracas of financing ideas, cloudy contracting practices, the rights and obligations of different levels of government, and competing perspectives on the correct balance between private industry and public control, emerged the first statewide solid waste management plan for New Jersey.[45] Financed through a federal grant enabled by the Solid Waste Disposal Act, an introductory letter from William V. Pye, president of the contractor-author Planners Associates, Inc. observes: "As New Jersey moves into the 1970s, the question of solid waste management must be given the attention it deserves within the overall environmental area. In this, the nation's most highly urbanized state, the question is a particularly urgent one. With the completion of this report, the State now becomes one of the first in the nation to publish a Solid Waste Management Plan."[46] The plan is positioned as a summarizing document, providing the "concepts and alternatives . . . the basic framework and guidelines for the subsequent solid waste management implementation programming that will take place at the State level . . . [and] be utilized by regional and county administrative units as an initial point for the more localized planning process."[47] Divided into nine chapters, the 1970 statewide plan begins with a handful of major findings and conclusions, serving to synthesize knowledge and perspectives about New Jersey's waste management system:

1   There are basically 567 independent waste management systems now
    operating in the State, far too many to function effectively according to
    most authorities on solid waste management.
2   The collection and disposal industry or system is too fragmented to invest
    in new technology...
4   ...in many areas, landfilling is used as a euphemism for stockpiling. That
    is, in the absence of any disposal or reclamation method, landfilling was
    practiced for no other reason than to provide an expedient storage place
    for refuse. Only vague plans for future development existed for many of
    these sites...

CONCLUSIONS [FROM THE 1970 STATEWIDE PLAN]
1   The current system of waste management is inadequate, and a perpetuation
    of current methods and facilities is undesirable...
2   There is no technological reason why we cannot dispose of solid waste with
    minimal environmental harm. The difficulty is one of application of known
    or experimental technology and methods.
3   The scale of current waste management systems is such that, with the
    exception of major central cities, few systems are large enough or well
    funded enough to engage in active experimentation. The large central cities,
    while of sufficient size... are confronted with other problems which to
    them are more pressing.
4   The major cost components of solid waste systems are, in order of magnitude,
    (1) hauling the waste... (2) collecting waste at pickup points and
    (3) the final disposal operations... [However] the disposal component...
    is the most significant component in terms of system performance.
6   Cost... is the most measurable aspect of a waste management system's
    performance; yet there are no standards as to how much waste collection
    should cost.[48]

Given the relative newness of waste management as a topic for statewide concern and coordination, the authors of the plan offer readers a crash course in
waste management, laying out definitions for solid waste and sketching the ways
in which "The current crisis in solid waste management has emerged before the
public suddenly" even as "conditions which led to the crisis had ... developed
over many years."[49] Accordingly, the plan locates the origins of the waste crisis
in New Jersey in the rise in standard of living and expansion of suburban lifestyles into previously rural landscapes. The plan points to the increasing costs
of waste management, both financially but also in terms of public health, deple-
tion of raw materials for manufacturing purposes, and damage to air and
water resources, as motivating factors for urgent action. In light of the findings
and conclusions offered in the opening pages of the document, the plan makes

the case for strong government involvement in addressing the waste management problem, and positions the state of New Jersey as the ideal designer and implementor of a comprehensive waste management plan.

The plan argues that the solid waste problem was previously confinable to the boundaries of each town and thus manageable by local governments alone. But waste management, like open space preservation, air pollution, and water quality, had become an issue crossing jurisdictional boundaries, "and the emerging recognition of these supra-municipal needs has led to acknowledgment of the necessary role of state government in . . . solid waste management."[50] To those ends the plan lays out six objectives for the state of New Jersey's involvement in solid waste:

1 To aid in providing in, and for, the State of New Jersey a system for the management of solid wastes, [which]

   a . . . is based on a rational planning program considering all relevant data;

   b . . . is economically feasible, though not necessarily least cost . . . ;

   c . . . provides for orderly and adequate collection of solid wastes;

   d . . . insures the continuing existence of a suitable and adequately distributed number of storage and disposal sites;

   e . . . considers all appropriate methods of collection, storage and disposal of solid wastes;

   f . . . provides the greatest possible compatibility with and the least possible disruption of both the state's urban population and natural resources; and

   g . . . is compatible with related Federal, state and local plans and programs.

2 To encourage, where appropriate, intra- and inter-governmental cooperation and coordination with respect to solid waste management.

3 To propose and encourage the adoption of appropriate Federal, state and local legislation in furtherance of these objectives.

4 To encourage the expenditures of Federal, state and local funds as required to meet these objectives and insure to the citizens of New Jersey an environment of suitable quality.

5 To establish appropriate standards for solid waste collection, storage and disposal within the state and insure compliance with those standards.

6 To inform the citizens of the state of the critical importance and value of an effective solid waste management program.[51]

The plan describes the federal role as supporting research and development for collection and disposal technologies, and especially, supporting nascent recycling practices and resource recovery technologies aimed at limiting the "unnecessary promiscuous use of the nation's resources," whether through direct subsidy or policy actions like creating recyclability standards for packaging.[52] In contrast to prevailing waste management practices of the twentieth century, the plan both minimizes the municipal role in waste management to simply collecting waste and undertaking inspection and code enforcement actions, and remains conspicuously silent on the role of private firms and contractors. In place of both, the plan proposes a greater state role in both collection and disposal, noting, for instance, "These objectives indicate the potential breadth of the appropriate role for state solid waste management, involving the collection and disposition of solid waste of all types throughout the state, either by direct state action, by regulation of municipal activities, or by state support of solid waste management activities at a level between the municipalities and the state."[53]

With statements like these, the 1970 statewide plan simultaneously reinforced and refuted the collective wisdom expressed in the documents presented to this point. For instance, the plan featured strong language highlighting the importance of coordinated action among New Jersey's municipalities, but hesitated about the possibility for voluntary collaboration among cities and towns; the plan encouraged greater investments in new collection practices and disposal technologies, but hardly acknowledged the private firms that would likely develop and operate improved waste management systems. On the one hand, plan authors likely knew that the variety of approaches to waste management already existing in New Jersey at the time of publication would have limited their ability to craft and present a universally acceptable, detailed, step-by-step pathway to the future of waste management anyway. On the other hand, approximately half of the statewide plan is dedicated to a methodical, quantitative analysis of population patterns, road and rail transportation networks, environmental characteristics, and solid waste generation statistics, designed to derive a network of "optimum" waste collection catchment areas and disposal sites that could be imposed on New Jersey cities and towns should lawmakers so choose.

The waste generation statistics were vitally important to the overall planning exercise, because the statewide model for waste management the authors developed hinged on the waste tonnage variables in conjunction with the population and transportation factors described above. Thus, there is extensive attention paid to quantifying waste volumes at the town scale, not only through corralling the existing data gathered by other state agencies in the late 1960s but also through projections developed by the plan authors for the year 1987. Careful explanation is offered for each category of waste (domestic, commercial, industrial, agricultural, and miscellaneous—"street refuse, dead animals, abandoned vehicles, fly-ash, incinerator residue, boiler slag, sewage treatment residue"[54]), along with demographic and socioeconomic context. At the same time, the authors are sharply critical of the waste management data available— "At this stage, the most significant conclusion of this investigation was that the data reported . . . was found to be too imprecise to be of use in projecting future levels of waste generation or . . . factors which were thought to influence waste generation."[55]

Their model for crafting alternative scenarios for New Jersey's waste management future rested on a derived "function relating waste density, average haul distance and total waste delivered to a point" in order to estimate the geographical boundaries of an "optimum collection district."[56] In other words, rather than attempting to resolve the political problem of communities (dis) agreeing to cooperate with one another in matters of collection and disposal, the authors created a system of waste collection areas based on how much it cost per mile to haul waste in different parts of the state and how much waste was generated there. The waste from each "collection district" was then directed toward a landfill, incinerator, or transfer station based on the historical disposal costs for that facility type and its distance from the collection district. Accordingly, all of the municipalities in the state were categorized by disposal facility type, becoming either "sanitary landfill" or "incinerator" communities based on the form of disposal that would be most feasible given site availability, settlement density, and estimated costs of disposal.[57] The abstract model led to two "theoretical alternatives": "total landfill system" and "total incineration system," wherein each technology was assumed to be the only one available for use; and the model applied fully using the actual waste generation data produced two "feasible alternatives": "maximum landfill strategy" and "maximum incineration strategy," each of which blended use of sanitary landfills and incinerators to emphasize one technology or the other.

Before presenting the results of each modeling approach, the plan offers an overview of disposal facilities in New Jersey at the end of the 1960s:

> With the exception of seven municipal incinerators, the only disposal method now employed in New Jersey is through sanitary landfill. The range of effectiveness of disposal techniques varies from unsatisfactory sanitary landfill (or open dump) to well-run sanitary landfill. In 1969, the State Health Department Survey of landfill operations reported the following:
>   Number of operating landfills   316
>   Number operating satisfactorily   32
>   Number operating unsatisfactorily   284
> It might be inferred from this study that the majority of landfill sites did not live up to the description of sanitary. A comparison of per ton landfilling costs vs. landfill size shows that the majority of landfill operating costs in New Jersey falls below the range recognized . . . as necessary for minimal operation.

The result of these circumstance is a bad name for landfill and an understandable reluctance of citizens to permit a sanitary landfill to operate in their vicinity.[58]

The plan theorizes that the maximum amount of land needed in New Jersey for landfills, for an estimated future population of ten million residents at projected growth rates of waste generation, would be just "11 square miles . . . 1/8 of one percent of the State's total area . . . or 1/4 of one percent of the State's open land in 1970," but notes that "despite the availability of potential land . . . lack of adequate landfill sites is still a real problem, not related to aggregate space needs but to quality of operation and compatibility with other land uses."[59]

With this caveat in mind, the plan describes the "total landfill system" and "total incineration system." Both of these represented theoretical maxima if either landfills or incinerators were mutually exclusive options for waste disposal in the state. The plan notes that neither approach is truly realistic, because there are parts of the state—especially the metropolitan counties in the northeast—that simply do not have sites available for large landfills; likewise, there are parts of the state—far northwestern New Jersey and South Jersey barring the northwestern half of Camden County—where land is relatively abundant and thus the additional costs of incineration would seem egregious. Accordingly, the plan focuses on two approaches to a blend of the disposal technologies, the "maximum landfill strategy" and the "maximum incineration strategy."[60] Maximum landfill

> is based on the assumption that, except for those areas in which the method is
> absolutely not feasible or desirable, sanitary landfill will remain the least
> expensive and most feasible form of disposal throughout most of New Jersey,
> and therefore a solid waste management system should utilize this method
> insofar as possible . . . A survey of currently available open land throughout the
> State's municipalities revealed that incineration would be necessary only in all
> of Hudson, Essex, Union, and Bergen counties and in portions of Passaic and
> Camden counties . . . For the State as a whole, it indicates a total of seven
> incinerator districts and thirty-two collection districts utilizing landfills.[61]

And likewise, maximum incineration

> assumes that landfill will eventually become an obsolete disposal method in all
> but the State's most rural areas, because of increasing development pressures
> and land costs in the growing urban and suburban counties . . . the approach in
> this alternative is to trade off efficiency and least cost in many of the districts
> served by incinerators in the short run for greater efficiency and modern
> facilities in the long run . . . The number of collection districts utilizing
> incineration would double from seven [in the maximum landfill strategy] to

fourteen in this alternative . . . if incineration is to be applied in those parts of the State undergoing rapid urbanization. This includes the balance of Passaic and Camden counties; all of Middlesex, Mercer, and Gloucester counties; and portions of Morris, Monmouth, Burlington, and Atlantic counties.[62]

Despite positive remarks about landfilling, the plan takes the position that waste disposal in New Jersey will inevitably trend toward incineration, describing maximum landfill "as one which is particularly adaptable as a short range strategy," but one which given New Jersey's rapid suburban development "may become difficult to maintain landfill operations . . . or conversely the waste densities may become sufficiently large to justify the use of incineration."[63] Similarly, the authors describe the maximum incineration strategy as "more adaptable in the long run, although it has less justification in the short run . . . this may be considered as a later stage of what could begin as the maximum landfill strategy . . . districts which start out utilizing landfill could later be consolidated into larger districts utilizing incineration."[64]

While the plan consistently reinforces the impression that the state alone has the depth of knowledge and expertise to develop an effective solution for waste management, in discussing the "feasible alternatives" the authors make clear that county boundaries are an ideal organizing principle around which to implement either maximum landfill or maximum incineration: "As can be shown . . . there are many equally feasible districting schemes possible in the State of New Jersey, based upon solid waste generation and facility location . . . it was shown that optimum sized districts were always multi-municipal. On a purely administrative basis, the multi-municipal unit of government, the county, could serve as a waste management administrative district. It can be seen that the basing of administrative districts on existing county lines, and the inclusion of one or more collection districts within each county, would not impose any basically uneconomical operating conditions upon any collection district."[65] Beyond simply observing that the multimunicipality waste collection districts derived by the modeling exercise were generally congruent with county boundaries in New Jersey, the plan authors made a more forceful case for placing county government at the center of waste management planning in the years ahead:

the Plan described herein is from the perspective of the entire State, utilizing the counties as sub-divisions thereof; it does not relate specifically to individual municipalities or economic establishments. The task of formulating a plan for solid waste management is viewed in broad terms as a State task and secondarily as a responsibility of each county . . .

However, the county has also emerged with a very important role as the key administrative unit and implementor of the State solid waste management system. In defining regionalized collection districts, county boundaries have been adhered to. This adherence to county lines in district formulation has been shown not to be a constraint in the optimization process. Within each county administrative unit, there emerged between one and five collection districts . . . They are not presented as rigid, final systems to be adhered to blindly. For example, the sub-county collection districts (which are configurations of municipalities) can and should be redefined at the county level where more thorough knowledge of local political, physical and other factors can be applied.[66]

In sum, the 1970 statewide plan and its methods were presented as logical, scientific, and, in contrast to much of the testimony received during the 1969 Special Legislative Commission hearings, apolitical. The quantitative functions developed by the authors produced a comprehensive collection and disposal strategy for New Jersey, without mentioning particular counties or towns by name and leaving specific decisions about implementation and facility siting to policymakers at the county and town level. Per the plan, "The use of this method indicates not the optimum location of waste disposal facilities, but the ideal spacing. Given a choice of feasible locations, several real alternatives may now be developed which fall within the ideal range as indicated."[67]

State- and county-level waste management plans would certainly have the greatest conceptual and practical impacts on New Jersey's waste management infrastructure, but in the late 1960s and early 1970s there were also a few super-regional plans that involved New Jersey, including one from the Metropolitan Regional Council of New Jersey, New York, and Connecticut (1971) and another from the "Tocks Island" Regional Advisory Council encompassing what is today known as the Delaware Water Gap National Recreation Area, at the intersection of northwestern New Jersey, east-central Pennsylvania, and southeastern New York state.[68] If state- and county-level waste management plans faced barriers to implementing their solutions, multistate regional waste management studies would be all but doomed to fail, from a practical stand-point, before the ink was even dry. The multijurisdictional nature of these regional projects means that the plans focus more on reporting data and developing abstract concepts for others to implement, than in developing immediately practicable solutions to any given problem.

As such, the Tocks Island study professes that "Much of the solid waste problem . . . is not technical but related to man himself [sic]. This report is more concerned with communicating to man about ways he can handle the solid waste problem better . . . all that may be needed is cooperation and respect: man's cooperation with his neighbor, and man's respect for his environment."[69] With this perspective at the forefront and echoing the 1968 New Jersey Solid Waste Program call for coordination of waste management planning with other types of planning exercises, the Tocks Island plan was oriented toward understanding how waste management needs could change in the face of population and tourism growth in the region, and evaluating different disposal technologies and their costs. The plan is interesting in that it identifies the many declining or abandoned strip mine sites in Pennsylvania in particular as catalyst sites for the development of large, inexpensive, regional landfills that could serve cities a considerable distance away.[70] The plan models several different scenarios based on highway or rail transport for waste, as well as varying degrees of cooperation among the multiple government jurisdictions, and concludes without advocating for any particular course of action.

The extent to which super-regional approaches to managing wastes penetrated thinking about garbage in New Jersey is not entirely clear. But by the turn of the 1970s the topic had gained considerable traction among all levels and components of government in the state. All of these early planning efforts, whether local, county, or statewide, reflected the reality that taking waste management seriously would be a complex matter involving the constituencies and finances of multiple levels of government. As the 1970s unfolded, attention turned toward addressing the concerns about New Jersey's waste management infrastructures that experts had identified in testimony before state government, and implementing ideas floated in documents like the 1970 statewide plan. Chief among these included strategies for determining appropriate prices for waste collection and disposal services, and resolving the questions over access to disposal sites that had been identified in the 1969 hearings before the Special Legislative Commission. The solution to these issues seemed to lay in finding ways of better coordinating New Jersey's towns, which as ever, remained on the front line of waste management and would bear the heaviest costs should any "crisis" in collection or disposal erupt. Circling all of these problems were questions about the role of private firms in handling the New Jersey's waste streams. While the administrative unit of the county seemed to be an attractive pathway for coordinating and centralizing solid waste activities, strategies for convincing—or compelling—county government to take the lead on waste management had yet to be devised.

3

Planning, Siting, Operating, and Financing Landfills

● ●

> This is a sovereign state, not a dumping
> ground, and this state government will
> act decisively to protect the health and
> welfare of its citizens.
> —Hon. Alfred N. Beadleston,
> New Jersey State Senate President,
> October 19, 1973

In July 1971 the New Jersey Department of Community Affairs (NJDCA) published in their statewide newsletter, *Community*, that the newly formed Hackensack Meadowlands Development Commission "has denied an application of the Municipal Sanitary Landfill Authority, a private solid waste disposal firm, to operate sanitary landfills on a total of 900 acres of meadowlands on different sites in the Kearny portion of the Meadowlands District." The Commission also passed regulations for solid waste management practices in the Meadowlands area aimed to "limit solid waste disposal to those sites already used for that purpose and rule out the use of new lands for garbage disposal."[1] The immediate cause for the denial, involving the same firm that had become the focus of Kearny mayor Joseph Healy's testimony during the 1969 Special Legislative Commission hearings, was that some 900 acres of the proposed landfill expansion site had been identified as either a prime target for wildlife and ecosystem preservation or deemed too close to residential areas of the town. In contrast to the widespread attitude held just years earlier, suddenly the Meadowlands area had become an important ecological site in the eyes of many state and local officials. As Edmund T. Hume, then chairperson of the Meadowlands Commission, noted: "Today's action, coupled with the Commission's denial on April 6 [1971] of an application to open a landfill site on 440 acres of adjoining land in Lyndhurst, and the adoption of the new regulations, saves from garbage disposal the entire Sawmill-Kingsland Marsh Conservation Area . . . It also preserves an additional 400 acres of tidally connected wetlands."[2] Hume argued that the denial of these applications in tandem with the conservation land use decisions, represented "the first step in the Commission's plan to terminate sanitary landfill within the Hackensack Meadowlands District and to provide modern facilities for the disposal and treatment of solid waste."[3]

Yet, as the unnamed author of the NJDCA newsletter article points out, "Under State law, the Meadowlands Commission is required to provide for the disposal of some 26,000 tons a week of solid waste in the District."[4] If the commission were to limit landfilling and other waste management operations in favor of environmental protection, where would all of the waste go? The question was clearly of importance to surrounding communities, too: as part of the formation of the Meadowlands area, more than one hundred communities in the surrounding counties were guaranteed waste disposal service at Meadowlands landfills. And, what would happen if landfills and dumps all across New Jersey shut their doors, not just those in the Meadowlands?

The 1970 statewide plan included a map showing all known waste disposal sites in New Jersey at the time of publication, including information about facility type (landfill or incinerator), ownership (public or private), and an estimate of the amount of waste received by each per day. The map shows more than 300 different disposal facilities in New Jersey. In 1985, when a draft update to the statewide plan was developed, that number had fallen to just fifty-nine.[5] In the process of implementing changes to New Jersey's waste infrastructure, why had so many landfills closed—and what would take their place?

Of course, many of the existing *landfills* in New Jersey were little more than dumping sites on the edge of town aimed at corralling trash into a single location, or perhaps originated as an effort to recover some value from a swamp, exhausted quarry, or otherwise *worthless* land held by a private owner. Since the 1950s, the State Sanitary Code had set rules for operation of dump sites on public health grounds, aiming to control odors and various disease vectors attracted to garbage, but with limited success: the 1970 Statewide plan reported that about 90 percent of the known landfills in the state operated afoul of state or local health rules.[6] Simultaneously, as new forms of environmental concern emerged in the United States during the 1960s,[7] laws were passed in many states to conserve natural resources and fragile ecosystems. In New Jersey, one of the major efforts was to establish the official Meadowlands area, but another was implementation of new wetlands and riparian protections throughout the state that limited landfill operations in coastal and marshy areas.[8] Taken together, as one planning study noted, the unsurprising result was that in New Jersey, "New landfill sites are extremely difficult to obtain. They are bitterly resisted by citizens in even moderately built-up areas, and are generally barred by environmental protection laws in undeveloped areas such as the wetlands. More remote sites have the added disadvantage of requiring greater hauling expenses."[9]

So just where would the garbage go, and how would it get there? This question has been at the heart of waste management policy not only in New Jersey but most states and localities in the United States. This chapter examines the state of New Jersey's efforts to answer this question, first through asserting legal authority to actually address the pending *trash crisis* and next through tackling the challenges of out-of-state waste being disposed in New Jersey landfills and dumps. The state also had to contend with persistent county opposition to meaningful action on comprehensive planning while developing a model for reliably financing new disposal facilities. It is important for readers to understand that these issues, while apparently obscure and removed from most folks' daily concerns, represent critical limiting factors for decisions that were taken in subsequent decades.

The Solid Waste Management Act and the Solid Waste Utilities Control Act

In conjunction with the operation and siting impacts of health codes and environmental protection laws, alongside publication of the 1970 statewide plan, the New Jersey legislature passed that same year the Solid Waste Management Act (SWMA). The SWMA assigned the newly formed New Jersey Department of Environmental Protection (NJDEP) powerful oversight and enforcement powers over the waste management industry. The SWMA shifted the jurisdiction of health and environmental dimensions of waste management to the NJDEP, and granted the agency authority over most aspects of both waste collection and disposal.

The 1970 SWMA authorized the NJDEP to seek injunctions and other legal relief against waste management disposal operations that did not meet new standards. In particular, the NJDEP developed rules around engineering standards and operating performance for disposal facilities in the state, including, for example, prescriptions on "permissible landfill configuration ... the periodic placement of cover material over exposed surfaces of solid waste and [a] mandate that new landfills provide gas venting and monitoring systems and groundwater monitoring systems," among other requirements.[10] Because the 1970 SWMA granted the NJDEP alone the authority to issue operating permits to all disposal facilities in the state, the operations of most of the landfills and incinerators in the state came under review—and several were closed for environmental reasons.[11]

Much of the information collected from the hearings, testimony, and research of the 1960s indicated that problems in waste collection also needed the attention of the state. As the New Jersey State Commission on Investigation reported in 1960, many waste management associations and professional groups:

> often bar new members unless they first receive approval from 75% of their existing members. The effect of these provisions and practices, of course, is to greatly discourage competition in the industry. By-law provisions encourage collusive bidding and preserve allocations of customers either by territory or on some other basis. The allocation of customers is perhaps the greatest vice in the industry. At present, there is no legislation in the State of New Jersey which prohibits garbage collectors from parceling out towns, areas or customers to one another. Unless this vice is checked, more and more municipalities will be faced with the situation where they receive only one bid for their waste collection contracts. It is a take-it-or-leave-it proposition in a situation where you can't leave it.[12]

In response, along with the 1970 SWMA, the New Jersey legislature passed the Solid Waste Utility Control Act (SWUCA). SWUCA assigned powers over waste collection pricing to the New Jersey Public Utility Commission, the same entity that regulated monopoly electricity providers in the state. This decision was an innovative one in the treatment of both the waste industry and the management of monopolies in the United States: as environmental attorney Dennis Krumholz noted in 1983, "By including under the jurisdiction of the Public Utility Commission great numbers of small and medium-sized businesses engaged in garbage collection and disposal, the Legislature departed to some extent from the traditional concept of a public utility which typically had been a monolithic entity."[13] The 1970 SWUCA was "intended to increase the economic fairness and efficiency of the solid waste industry by establishing just and reasonable rates for collection and disposal services and by eliminating abuses, such as favoritism, corruption and bid-rigging, which have plagued the industry."[14] Key provisions of SWUCA were to grant the Public Utility Commission (today known as the Board of Public Utilities, BPU) the power to certify and revoke certification of individuals "based on experience, training, or education," to designate franchise areas for collection, to regulate and adjust rates for collection and disposal, and to ensure reasonably competitive bidding for contracts.[15] The "split authority" over solid waste between the NJDEP and the Public Utility Commission recognized the complexity of the waste management industry: both collection and disposal lie at the intersection of environmental, economic, and political concern.

The interjection of state authority into the waste problem was welcomed by most, but also invited legal challenges. For example, the *Ringlieb v. Township of Parsippany-Troy Hills* case—which went all the way to the Supreme Court of New Jersey—centered on whether municipalities could still issue their own regulations aiming to shape waste disposal, or whether the new SWMA and SWUCA laws preempted local rules.[16] Here, the supreme court found that with the two laws, the state of New Jersey intended to create a "comprehensive plan on the part of the State to control all facets of this industry," and that the passage of new municipal ordinances would not only impose "double effort" on waste disposal facilities, but also that "the conflicting ordinances and requirements of the separate municipalities would bring to a complete halt the sanitary landfill operations in this state, the refuse disposal business, all to the detriment of the general health of the general public."[17] In other words, the court held that with the passage of SWMA and SWUCA, the state of New Jersey secured exclusive jurisdiction over waste management, and concluded that municipalities could not create their own regulations in this area. Furthermore, the court concluded that towns had no right to legal remedy or even administrative hearings to oppose the operations of a landfill within their boundaries, if the facility was found to have been properly certified by the NJDEP and Public Utility Commission.[18]

By 1975 it was clear that the state of New Jersey had centralized power over waste management, and undoubtedly many unhealthy, environmentally problematic landfill operations were closed down during the 1970s and 1980s. It is equally clear that the state would no longer permit establishment of new disposal sites and especially landfills in environmentally sensitive areas. While these developments were unambiguously positive from the perspective of ecosystem and natural resource conservation, they surely contributed to the rapid diminishment of waste disposal capacity. The legislature, NJDEP, and Public Utilities Commission had developed and implemented effective oversight tools and enforcement actions, but not yet answered the question of what to do with the increasing amounts of garbage produced by New Jerseyans and their businesses.

Thus, for much of the 1970s and 1980s, one word would come to be used more than any other to describe waste management in New Jersey: *crisis*. The crisis narrative centered on disposal capacity, and in particular, the projection that "dedicated landfill acreage will be exhausted in the State within the next 10 or 11 years," and that "landfills serving 5 of New Jersey's 21 counties—Bergen, Essex, Hudson, Passaic and Union—which produce 48 percent of the State's annual volume of solid waste will be exhausted during 1975."[19] Nearly every observer reached the same conclusion. As summarized in one 1972 report on municipal and county cooperation,

> Solid waste collection, processing and disposal problems in New Jersey are approaching a critical stage, while our capabilities of dealing with these problems fail to measure up to the task. The data developed in this study reveal that during 1971 a total 7.1 million tons of solid waste was produced in the State and, if recent trends continue, this amount . . . will increase to over 22 million tons annually by the year 2000. This staggering amount of waste, confronting an inadequate management system, threatens New Jerseys' environment and the health and welfare of its citizens. Solid waste problems are reaching crisis proportions in virtually every area of the State, yet the critical need for effective state-wide and regional approaches to waste management still goes unmet.[20]

The lack of industry and facility oversight, and seeming inability of the state to regulate waste, were solved at least conceptually with passage of the SWMA and SWUCA in 1970. But neither of these laws was able to fully address the crisis in disposal capacity itself: even if state agencies could set the rules for waste disposal, they were so far unable to answer the question of where to put the stuff. Tackling the disposal crisis demanded efforts in three further areas: managing waste originating out of state but being disposed in New Jersey; overcoming county inertia and even resistance to waste planning and especially facility siting; and directing flows of wastes in an economical fashion while also resolving the murkiness around waste collection practices and contracts.

Blocking the Flow of Out-of-State Wastes(?)

On October 19, 1973, state senate president Alfred N. Beadleston proclaimed:

> New Jersey cannot continue to tolerate the grave environmental dangers being created by the dumping of garbage in New Jersey from New York or any other state. This is a sovereign state, not a dumping ground, and this state government will act decisively to protect the health and welfare of its citizens. Not only does out-of-state dumping pose immediate environmental and health problems, but over the long run, the people of New Jersey will be forced to pay much higher costs for waste removal because available landfill dumping areas are filled up.[21]

Subsequently Richard Sullivan, then commissioner of the NJDEP, was directed by the office of Governor William Cahill to restrict or halt entirely the flows of waste coming from New York and Pennsylvania, and in particular the cities of Philadelphia and New York City. There was a particular focus to implement rules banning out-of-state waste disposal in the Meadowlands, but also a parallel legislative effort to permanently restrict inflows of trash anywhere else in New Jersey. A *New York Times* article about the ban estimated that more than a combined 3,000 truckloads of waste from New York City and Philadelphia entered the state each week, amounting to about 1.5 million tons of trash each year. In other words, approximately one fifth—21 percent to be precise—of all waste disposed in New Jersey's landfills in 1973 came from outside state borders. One landfill in Mount Holly, a semirural community in Burlington County, in southern New Jersey, was estimated to receive one-third of the garbage generated in Philadelphia. Charles Gingich, chief environmental specialist of the Bureau of Solid Waste Management in the NJDEP, noted in the same article that citizens in Edison Township, Madison Township, and Plainfield, had all lodged complaints with the bureau that "out-of-state truckers" were rapidly depleting "landfills established for their own solid waste disposal needs."[22]

Of greatest concern was the projection that the landfill capacity of the Meadowlands area, recipients of nearly half of the state's garbage, would be completely exhausted by 1975. In response, the Hackensack Meadowlands Development Commission (HMDC) along with NJDEP established new rules in 1973, explicitly stating that:

> No solid waste originating or collected outside of the territorial jurisdiction of New Jersey shall be disposed of or treated within the Hackensack Meadowlands District. No sanitary landfill operator shall accept for disposal, at a sanitary landfill within the Hackensack Meadowlands District, any solid waste originating or collected outside of the territorial limits of New Jersey.
>
> All operators of sanitary landfills within the Hackensack Meadowlands District shall submit to the Commission . . . a certification stating that no solid waste originating or collected outside of the territorial limits of New Jersey will be accepted for disposal or treatment.[23]

The commission estimated that such a ban would extend the life of the landfills in the Meadowlands area, affording them time to develop alternative disposal facilities like an incinerator and a baling facility that could compress solid waste into large blocks that could be used for some construction purposes. At the same time, the state legislature passed a law banning disposal of out-of-state waste (applying to all materials except "garbage to be fed to swine") at sites anywhere in New Jersey. Foundational to both the Meadowlands ban and the statewide ban, codified as the Waste Control Act of 1973, was the explicit protection of the environment. The preamble to the text of that law reads: "The Legislature finds and determines . . . that the treatment and disposal of these wastes poses a threat to the quality of the environment, that the quality of New

Jersey's environment is being threatened by the treatment and disposal of wastes generated or collected outside the State, and that this hazard can be reduced by the adoption of regulations governing this practice."[24]

Almost immediately thereafter, the HMDC together with NJDEP sought an injunction against the Municipal Sanitary Landfill Authority (MSLA) of Kearny—the same firm at the heart of the controversy over municipal bidding just six years earlier. The facility operators insisted on their right to continue to accept waste from New York City and surrounding suburbs like Yonkers.[25] MSLA contended that the regulations banning the disposal of out-of-state waste in Meadowlands landfills were invalid for three reasons: first, because the ban on disposal in the Meadowlands was "arbitrary and unreasonable"; second, that the NJDEP had violated due process in developing and enacting the regulations; and third, that a ban on disposing in New Jersey waste originating in another state violated the United States Constitution's "Commerce Clause" granting oversight of trade and economic activity involving several states to the federal government alone.

Brought first before the chancery division of the Superior Court of New Jersey in 1974, Judge Sidney M. Schreiber found MSLA's first two claims to be baseless. In particular, the court found that the state of New Jersey through the HMDC and NJDEP had every right to regulate waste disposal in the Meadowlands area because the language of the enabling legislation made clear that both HMDC and NJDEP were tasked with understanding and managing the total amount of garbage disposed in the state, regardless of its origins, in the name of environmental protection. In other words, the HMDC and NJDEP surely did have the power to monitor and impose limits on waste disposal in order to conserve natural resources (like land) and limit pollution.

However, Judge Schreiber refuted HMDC's and NJDEP's position that a ban on "waste," as a set of materials having no economic value, could not in concept violate the Commerce Clause. In contrast, Schreiber affirmed the position of the MSLA that *some* types of waste clearly did have economic value—as building materials, recyclables, or as fuel for incinerators. While Schreiber acknowledged a number of cases that affirmed the "police power" of state governments and agencies to limit or prohibit the flow of materials that could harm humans—diseased animals, tainted produce, contaminated food products, etc.—there were some important distinctions to be drawn regarding the flow of waste from one state to another. He wrote in his opinion that

> the regulations and statute in question are clearly within the authority of the State's police power to conserve its landfill sites. It has been recognized that the dumping of garbage in a landfill is an unattractive and undesirable use of land and that it creates potential health and fire hazards . . . In the exercise of its police powers, a State may exclude from its territory, or prohibit the sale

therein of any articles which, in its judgment, fairly exercised, are prejudicial to the health or which would endanger the lives or property of its people. But if the State, under the guise of exerting its police powers, should make such exclusion or prohibition applicable solely to articles, of that kind, that may be produced or manufactured in other States, the courts would find no difficulty in holding such legislation to be in conflict with the Constitution of the United States.

... DEP and HMDC rationalize that discrimination is absent because the regulations do not prefer New Jersey over out-of-state landfill operators or garbage haulers. The difficulty with that rationale is that it is the discrimination based on the origin of the refuse, not the residence of the owner or carrier ... Transportation of the refuse, whether carried by New Jerseyans or nonresidents, in interstate commerce is directly affected by the ban. Although the State's objective in attempting to conserve a local natural resource for local needs is a proper police power purpose, that cannot be accomplished by discrimination based on the source of the refuse.

... The Commissioner of Environmental Protection and the Hackensack Meadowlands Development Commission adopted the regulations in question to expand the life of the landfill sites for garbage collection in the Hackensack Meadowlands District from 3 to 3 1/2 years. To attempt to accomplish that end by discriminating against sources of refuse solely because they originate or are collected outside the State under the circumstances here contravenes Article I, Section 8, Clause 3 of the U.S. Constitution. The regulations are invalid.[26]

HMDC and NJDEP appealed, and a year later the case was heard before the Supreme Court of New Jersey. By the time a decision was reached on the appeal in 1975, the city of Philadelphia and a number of landfill operators in southern New Jersey had joined the case in opposition to the ban on out-of-state disposal.[27] Here, the matter was solely whether the state of New Jersey and its agencies had violated the Commerce Clause of the U.S. Constitution in establishing the ban on out-of-state wastes. The court determined that it was most likely true that the ban *did violate* the Commerce Clause in a strict sense, but that this violation was not proportionally greater than the environmental benefits that could be achieved. In other words, the economic value of the commerce of disposing out-of-state wastes in New Jersey was outweighed by the environmental protection values of the ban, limiting potentially dangerous materials from harming state resources and preserving comparatively untouched ecosystems and land. Justice Worrall F. Mountain, who wrote the opinion for the case, observed that "the state has not only a right to protect its own resources, but also the duty to do so, in the interest of its citizens as well as others" and that the majority of environmental, health, and consumer protection laws and rules inherently regulate interstate commerce to some degree.[28]

Furthermore, the court noted that all of the defendant landfill operators had alternative customers—that they did not deal only with out-of-state haulers as their business model—and that the cities of Yonkers and Philadelphia each had alternative disposal options nearly comparable in price. The opinion of the court was that the ban did not stop "foreign" (meaning, non-New Jersey) companies from operating waste collection or disposal businesses in the state or even limit the transportation of waste materials through the state, but that it only limited the particular practice of disposing in New Jersey landfills waste collected outside the state. For these reasons, the court reversed or vacated all of the lower courts' decisions. The unanimous decision, reported by Justice Mountain, noted that "We do not mean to be understood as saying that every assertion of a state's quasi-sovereign right to protect its environment, ecological values and natural and human resources will be sustained regardless of the impact upon interstate commerce. Clearly this is not so. But where the effect upon trade and commerce is relatively slight, as is here the case, and where at the same time the values sought to be protected by the state legislation are as crucial to the welfare of its citizens as is here true, we have no hesitancy in sustaining the state action."[29]

The 1975 Supreme Court of New Jersey decision left open several avenues for appeal. Most directly, the court acknowledged the reality that the ban likely did violate the Commerce Clause in a strict sense, but not in a material way since the negative commercial impacts were minor compared to the environmental benefits. The cities and landfill operators would clearly disagree with this assessment. But, more importantly, in its opinion the court pointed to another case about waste management from the 1960s decided by the Supreme Court of the United States.[30] *United States v. Pennsylvania Refuse Removal Association* was an antitrust case brought against a group of waste collection firms who decided to dispose of their waste in New Jersey, refusing to consider disposal facilities in Pennsylvania. Critically in this case (discussed in greater detail later on), the Court found that the "commerce" in question was not in fact the garbage itself, but rather the *service* of collection and disposal.

This distinction would prove vital. By 1978, attorneys representing the city of Philadelphia had taken the matter all the way to the Supreme Court of the United States. The proceedings suggest that in their arguments, the parties to the suit focused on the balance between economics and environmental protection, and whether one could outweigh another in justifying New Jersey's ban on disposal of wastes originating in other states. That is to say, the arguments heard by the Court focused on attempts to prove, or disprove, the intentions of the New Jersey state legislature in passing the Waste Control Act in the first place:

> Based on these findings, the [Supreme Court of NJ] concluded that ch. 363 [the Waste Control Act] was designed to protect not the State's economy, but its environment, and that its substantial benefits outweigh its "slight" burden on interstate commerce . . . The appellants [City of Philadelphia et al.] strenuously contend that ch. 363 "while outwardly cloaked in the currently fashionable garb of environmental protection is actually no more than a legislative effort to suppress competition and stabilize the cost of solid waste disposal for New Jersey residents." They cite passages of legislative history suggesting that the problem addressed by ch. 363 is primarily financial: stemming the flow of out-of-state waste into certain landfill sites will extend their lives, thus delaying the day when New Jersey cities must transport their waste to more distant and expensive sites.
>
> The appellees [HMDC / NJDEP / State of New Jersey], on the other hand, deny that ch. 363 was motivated by financial concerns or economic protectionism. In the words of their brief, "no New Jersey commercial interests stand to gain advantage over competitors from outside the state as a result of the ban on dumping out-of-state waste." Noting that New Jersey landfill operators are among the plaintiffs, the appellee's brief argues that "the complaint is not that New Jersey has forged an economic preference for its own commercial interests, but rather that it has denied a small group of its entrepreneurs an economic opportunity to traffic in waste, in order to protect the health, safety and welfare of the citizenry at large.[31]

However, Justice Potter Stewart in delivering the majority opinion, noted that from the Court's perspective the rationale for the ban was irrelevant. Unless a state could prove that something inherent to the material itself was problematic—not simply its origin—there could be no grounds for a state limiting flows of commerce. This distinction demands that waste disposal be understood differently from instances of quarantine prohibiting diseased livestock or contaminated food from entering a state. And regarding this precise matter, he wrote,

> The New Jersey statute is not such a quarantine law. There has been no claim here that the very movement of waste into or through New Jersey endangers health, or that waste must be disposed of as soon and as close to its point of generation as possible. The harms caused by waste are said to arise after its disposal in landfill sites, and, at that point, as New Jersey concedes, there is no basis to distinguish out-of-state waste from domestic waste. If one is inherently harmful, so is the other. Yet New Jersey has banned the former, while leaving its landfill sites open to the latter. The New Jersey law blocks the importation of waste in an obvious effort to saddle those outside the State with the entire

burden of slowing the flow of refuse into New Jersey's remaining landfill sites. That legislative effort is clearly impermissible under the Commerce Clause of the Constitution.

Today, cities in Pennsylvania and New York find it expedient or necessary to send their waste into New Jersey for disposal, and New Jersey claims the right to close its borders to such traffic. Tomorrow, cities in New Jersey may find it expedient or necessary to send their waste into Pennsylvania or New York for disposal, and those States might then claim the right to close their borders. The Commerce Clause will protect New Jersey in the future, just as it protects her neighbors now, from efforts by one State to isolate itself in the stream of interstate commerce from a problem shared by all.[32]

While the final words of the majority opinion were highly prescient—"Tomorrow, cities in New Jersey may find it expedient or necessary to send their waste into Pennsylvania or New York for disposal"—the decision to strike down New Jersey's ban on disposal of out-of-state wastes was not unanimous. Justices William Rehnquist and Warren E. Burger filed a dissenting opinion, focusing on the distinction the majority opinion had drawn between cases of quarantine and apparently less-hazardous trash. Justice Rehnquist pointed to the litany of health and environmental problems arising from improperly managed sanitary landfills, such as fire, attraction of disease vectors, or the pollution of ground and surface waters. The dissent continues

The question presented in this case is whether New Jersey must also continue to receive and dispose of solid waste from neighboring States, even though these will inexorably increase the health problems discussed above . . . The physical fact of life that New Jersey must somehow dispose of its own noxious items does not mean that it must serve as a depository for those of every other State. Similarly, New Jersey should be free under our past precedents to prohibit the importation of solid waste because of the health and safety problems that such waste poses to its citizens. The fact that New Jersey continues to, and indeed must continue to, dispose of its own solid waste does not mean that New Jersey may not prohibit the importation of even more solid waste into the State. I simply see no way to distinguish solid waste, on the record of this case, from germ-infected rags, diseased meat, and other noxious items . . .

Even if the Court is correct in its characterization of New Jersey's concerns, I do not see why a State may ban the importation of items whose movement risks contagion, but cannot ban the importation of items which, although they may be transported into the State without undue hazard, will then simply pile up in an ever increasing danger to the public's health and safety. The Commerce Clause was not drawn with a view to having the validity of state laws turn on such pointless distinctions . . .

... the Court implies that the challenged laws must be invalidated because New Jersey has left its landfills open to domestic waste ... The fact that New Jersey has left its landfill sites open for domestic waste does not, of course, mean that solid waste is not innately harmful. Nor does it mean that New Jersey prohibits importation of solid waste for reasons other than the health and safety of its population. New Jersey must, out of sheer necessity, treat and dispose of its solid waste in some fashion, just as it must treat New Jersey cattle suffering from hoof-and-mouth disease. It does not follow that New Jersey must, under the Commerce Clause, accept solid waste or diseased cattle from outside its borders, and thereby exacerbate its problems.[33]

Even if the ban on out-of-state trash in New Jersey landfills had been upheld by the Supreme Court of the United States, it was clear that blocking New York City's or Philadelphia's garbage could not fully resolve the waste crisis. At most, restricting these waste flows—about 20 percent of all the waste disposed in the state by the late 1970s—would have added only a few years to the working lives of New Jersey's busiest landfills. Responsibility for solving the waste crisis would still fall largely on the shoulders of state and, increasingly, county officials.

Disposal Facility Planning, Siting, and Financing

In 1972, William V. Musto, chairperson of the County and Municipal Government Study Commission, released a report analyzing the political dimensions of waste management in New Jersey and in particular, the relationship between state, county, and local officials as it related to handling trash. The "Musto Commission," whose membership was comprised of representatives from the state senate and assembly as well as representatives of various county and municipal governments, articulated in their report *Solid Waste: A Coordinated Approach* much more forcefully what had been an undercurrent in the 1970 statewide plan: that counties must begin to take the lead in waste management in New Jersey.

To date, only one New Jersey county [Bergen] has become involved in sanitary landfill operations ... Municipal officials in virtually every county have sought assistance from county officials in dealing with solid waste disposal. The county appears to be a logical level of government to plan and develop regional sanitary landfills, since:
- the county has a wider choice of possible sites than a municipality;
- it would be cheaper for the county to plan and/or operate one or two large landfills than several smaller ones, if the large landfills were centrally located;
- the survey, design, and planning work for one or two large county-wide landfills would also be cheaper;

- coordination of pollution monitoring and abatement activities would be easier; and
- the county has a broader tax base from which to raise revenues for the purchase of land and operation of the landfill.[34]

Following the 1970 statewide plan, the Musto Commission report argues for regional solid waste collection and processing districts, which could expediently match county boundaries. The report optimistically observes that attitudes toward regionalization were evolving, with "the climate of opinion among local officials . . . shifting strongly in favor of larger service districts."[35] Despite this newfound enthusiasm, "there have been few successful attempts to deal with comprehensive solid waste problems on an inter-municipal or county wide basis."[36] The report continues, "Failure to develop regional facilities can be attributed to the following factors 1) lack of planning, 2) inability to implement the plan once it is developed (primarily because of failure to agree on a site for the facility), 3) inadequate public understanding of the problem and 4) lack of funding. To date there has been almost a complete lack of planning at the inter-municipal or county level for the implementation of regional processing or disposal facilities."[37]

By the end of 1972, just six of New Jersey's twenty-one counties had produced even preliminary solid waste management plans: Bergen, Burlington, Camden, Essex, Mercer, and Monmouth. These early plans did not typically take strong positions on siting new disposal facilities, or in some cases, even suggest possible locations. Instead, as the discussion of Camden County's plan in the previous chapter showed, these documents presented the status of waste management in the county through facts, figures, maps, and charts. Earlier uncertainty about counties' ability to acquire a site (even if they had identified a suitable one and encountered zero public opposition) or questions about their ability to finance and operate a disposal facility were eliminated by 1972. The County Solid Waste Disposal Financing Law as passed in late 1970 provided county governments all of the authority they would need to develop and operate landfills or incinerators.[38] Yet, the Musto Commission report noted that "only a few" New Jersey counties had taken advantage of this new legal authority.[39]

County plans had so far avoided the thorny question of new facility siting for simple reasons: "The site selection impasse, the [Musto Commission] found, was bound up with citizen hostility toward the idea of being a neighbor to a solid waste facility, particularly a regional one. No one wants to live next door to this type of facility."[40] Counties that had engaged with the planning process, according to the Commission Report, "have had their initiative rewarded by bitter arguments over site location."[41] The 1970 SWMA would require changes in order to have counties take seriously their obligation to plan for garbage collection and disposal. In 1975, the legislature took action to amend the 1970 SWMA, granting the NJDEP important powers while also introducing new concepts to broaden the scope of stakeholder participation in waste management planning and even the concept of waste management itself.

The 1975 Solid Waste Management Act Amendments (326 amendments, in reference to the chapter of the 1975 laws in which the amendments appear) introduced several measures to spur county-level waste planning. The most direct was to designate each county, along with the Hackensack Meadowlands area, a "solid waste management district," and grant to county government the express authority to develop waste management plans for their own district or in conjunction with other waste management districts.[42] Subsequently, the 326 amendments demanded that each waste management district produce a plan for handling trash around the end of the 1979. This plan, once approved by the NJDEP, would be updated at least every ten years though amendments could be submitted in the meantime. The new county plans, per the text of the law, were required to consider:

A1 An inventory of the sources, composition, and quantity of solid waste generated within the solid waste management district . . . ;

A2 Projections of the amounts and composition of solid waste to be generated within the district in each of the 10 years following the year in which the report is prepared . . . ;

A3 An inventory and appraisal, including the identity, location, and life expectancy, of all solid waste facilities within the solid waste management district . . . and the identity of every person engaging in solid waste collection or disposal within the district;

A4 An analysis of existing solid waste collection systems and transportation routes within the solid waste management district.

B1 The designation of a department, unit or committee . . . to supervise the implementation of the solid waste management plan . . . ;

B2 A statement of the solid waste disposal strategy to be applied in the solid waste management district . . . and a plan for using terminated landfill disposal sites, if any, in the solid waste management district;

B3 A site plan, which shall include all existing solid waste facilities located within the solid waste management district, provided that they are operated and maintained in accordance with all applicable health and environmental standards, and sufficient additional available suitable sites to provide solid waste facilities to treat and dispose of the actual and projected amounts of solid waste contained in the report accompanying the plan;

B4 A survey of proposed collection districts and transportation routes . . . ;

B5 The procedures for coordinating all activities related to the collection and disposal of solid waste by every person engaging in such process within the solid waste management district . . . ; and the procedures for furnishing the solid waste facilities contained in the solid waste management plan; and

B6 The method or methods of financing solid waste management in the . . . district pursuant to the solid waste management plan.[43]

Where counties had hesitated in the past to consider new sites for landfills or other disposal facilities, the 326 amendments now forced them to meet this obligation. In their plans, not only would counties have to account for current waste volumes and estimate how they might change ten years into the future, but they would have to match these waste flows to an equal amount of disposal capacity within their borders. The 326 amendments contained provisions that if a county did not identify and begin development of new disposal sites within its own border—whether due to geographic or ecological constraints or pure opposition to the planning process—that it could seek to develop a disposal facility in another county or in conjunction with another county so long as both waste management plans reflected the arrangement. If none of those things happened, the 326 amendments granted NJDEP itself the authority to identify potential disposal sites and eventually force county government to use the sites NJDEP selected. Similarly, the NJDEP was granted the authority to create a plan for any county that refused to engage with the planning process or that was unable to have the plan approved by the county's elected officials.

Alongside the new solid waste management districts, the 326 amendments extended additional powers to the NJDEP itself. In particular, the laws granted NJDEP express authority to create a statewide waste management plan and to supervise and review the planning process of the counties in light of that plan. The laws elevated the statewide plan to official policy for waste management in New Jersey. For example,

It is the policy of this State to . . . establish a meaningful and responsible role for the State in the solution of solid waste problems by granting the Department of Environmental Protection . . . the power, not only to regulate and supervise all solid waste collection and disposal facilities and operations . . . but also to develop . . . criteria and procedures to assure the orderly preparation and evaluation of the solid waste management plans developed by every solid waste management district, and to approve, modify, or reject such solid waste management plans on the basis of their conformity with such objectives . . . to develop and implement such a plan where none is approved or forthcoming from any solid waste management district, to arbitrate disputes between solid waste management districts . . . all in the manner and extent hereinafter provided.[44]

In this way, the 326 Amendments demanded that counties propose waste solutions in response to their local setting but reserved for NJDEP the final authority over whether or not county plans were meeting statewide objectives.

It is interesting to consider the formation of the Hackensack Meadowlands area a few years prior to the 326 amendments in the context of these ideas. There should be no doubt that the region was designated a special planning district and land use management area for substantial environmental protection reasons. But the formation of the Meadowlands can also be read as a state-led demonstration project of multimunicipal cooperation around waste management. It had been evident for years that leaving Meadowlands towns to independently manage landfills and monitor contractor-operators was inefficient and prone to cloudy business arrangements. As the 1972 Musto Commission report observed, "the responsibility for providing landfill facilities is fragmented among many local governments and private landfill operators who simply lack the mandate and scope of operations to provide any guarantee that sufficient and environmentally acceptable landfill facilities will continue to be available . . . Recently there has been a trend towards an increased State role in the disposal of solid waste. Legislation creating the Hackensack Meadowlands Development Commission mandates to the Commission the responsibility of providing disposal facilities for 118 of New Jersey's municipalities."[45] Thus the approach to planning, and the strategies for waste management developed by the Meadowlands commission foreshadowed some of the content of the 326 amendments. The state showed how a regional approach to waste management could work in practice, while implicitly articulating how the state and NJDEP might impose a waste management plan on a county that could not or would not voluntarily develop one of their own. It is perhaps significant in this way that the Meadowlands is such an ecologically sensitive region, since many of the *outs* for counties in designating new disposal facility sites in the 326 amendments are premised on ecological criteria. That is to say, there are probably no places in New Jersey, from an ecological perspective, less suitable for waste disposal than the Meadowlands—yet if a waste plan could be developed and implemented there, surely waste plans could be developed and implemented everywhere else in the state.

Along with the mandate for counties to plan for waste management, the 326 amendments created opportunities and obligations for greater public participation in the waste management planning process. In the words of the legislation, the amendments were intended to "provide citizens and municipalities with opportunities to contribute to the development and implementation of solid waste management plans by requiring public hearings prior to their adoption and by the creation of advisory solid waste councils."[46]

The first of these advisory councils was created within the NJDEP. The Advisory Council on Solid Waste Management included the president of the Board of Public Utility Commissioners, the Commissioner of Community Affairs, the Commissioner of Health, and the Secretary of Agriculture, along with seven members of the public. Per the legislation, three of the public members "shall be actively engaged in the management of either solid waste collection or solid waste disposal, or both," and the other four "shall be representing the general public to be appointed by the Governor with the advice and consent of the Senate."[47] This body would study solid waste problems and emerging technological solutions, and "from time to time submit to the [NJDEP] commissioner any recommendations which it deems necessary for the proper conduct and improvement of solid waste programs and solid waste management plans."[48] This council met during the 1970s and 1980s, but with little impact on the county planning and landfill siting process; the body played a more substantial role around the development of recycling programs and infrastructure in New Jersey (discussed in the next chapter).

Second, and more significantly, the 326 amendments outlined a particular process counties would follow in developing and formally adopting their waste management plans. One provision of this process was that counties create a waste advisory council similar in scope and function to the Advisory Council on Solid Waste Management housed within the NJDEP. The county councils, however, would be comprised of "municipal mayors or their designees, persons engaged in the collection or disposal of solid waste, and environmentalists."[49] The amendments also noted that county waste management plans could not progress through the approval process without consultation from the county advisory council. Another provision of the county plan approval process was that county officials hold a public meeting "for the purpose of hearing persons interested in, or who would be affected by, the adoption of the solid waste management plan ... and who are in favor of or are opposed to such adoption."[50] The NJDEP reserved the right to hold additional public hearings in the event of a county failing to do so, or failing to adopt a waste management plan at all.

While requiring counties to act on waste management planning and creating new opportunities for public involvement in the planning process were key components of the 326 amendments, the law did contain other important provisions which are discussed in greater detail later in this chapter and in the next. In particular, the 326 amendments formally labeled solid waste collection and disposal as a "utility" subject to oversight from the Board of Public Utility Commissioners (PUC) and extended to PUC the authority to designate waste collection territories or "franchises" subject to utility-style regulated pricing.

The amendments extended a number of tax benefits (mostly payable by county governments) to towns hosting disposal facilities and authorized host towns to receive a 25 percent (maximum) discount on rates set by the PUC, and created a "Solid Waste Management Research and Development Fund" as well as a grant program to support experimental waste collection and disposal technologies. Finally, and for the first time, the amendments defined the terms "resource recovery" and "recycling facility" as components of the waste management arsenal. In fact, the 326 amendments included the formal renaming of the 1970 Solid Waste Management Act to include the words "resource recovery"[51] and included in the preamble to the legislation that "the Legislature, therefore, declares that it is the policy of this State to ... Encourage resource recovery through the development of systems to collect, separate, recycle and recover metals, glass, paper and other materials of value for reuse or for energy production."

With the 326 amendments the state of New Jersey established a clear hierarchy for waste management planning: NJDEP would set the overall approach and objectives, while counties would interpret those goals in their own specific contexts. If counties could not, or more likely, would not, comply, then the NJDEP would initiate a legal process to impose a waste management plan for that county. All of these efforts were intended to site and implement new disposal facilities and ease the landfill crisis. However, planning for waste management in many cases ended up being a totally different pursuit than actually developing disposal infrastructure. In fact, the planning process imposed by the 326 amendments precipitated a fascinating set of attitudes toward waste disposal as counties demonstrated varying degrees of commitment to solving the disposal problems within their own borders.

For example, Burlington County—a very large, mostly rural county in southern New Jersey—drafted plans for an extensive waste management infrastructure, ranging from strategies for collecting waste to development of a county-owned landfill and various recycling facilities.[52] Similarly, Cape May County in the far south of New Jersey actually succeeded in planning for, designing, and building a landfill by 1984.[53] However, proactive county participants in waste management planning were the exception rather than the rule. In fact, by 1985 the NJDEP determined that "the lack of progress in some districts in implementing Chapter 326 has been used to provide a basis for legal action against districts that have clearly failed to implement their Chapter 326 mandates ... 15 districts had failed to develop and/or implement plans adequate to meet their disposal needs and responsibilities delegated to them under the Solid Waste Management Act."[54] That is to say, between the effective date of the 326 amendments and 1985—about seven years—fifteen of New Jersey's twenty-one counties had either been sued by or pressed into some sort of consent agreement with the NJDEP over various deficiencies in their proposed waste management plans.

It was clear that the NJDEP was willing to back up the planning provisions of the 326 amendments with legal and administrative action. For example, Camden County (one of the fifteen noncompliant counties) in their "326 plan" identified six different resource recovery projects (recycling, incineration with energy recovery, or composting) as potential pathways toward security in waste disposal capacity. But by 1985 all but two of the resource recovery projects had fallen through completely, and the county had let its agreement with Burlington County to use their landfills lapse. The county was sending the majority of the waste generated by its municipalities to Kinsley's Landfill in neighboring Gloucester County. Despite eventually signing an agreement with Gloucester County over this arrangement—seemingly designating this course of action as *the* waste management plan for the foreseeable future—the NJDEP was not satisfied. The agency secured a preliminary injunction requiring Camden County to site a new landfill by spring of 1985, over fears that Kinsley's Landfill would reach its capacity before 1990. County officials selected a site in the town of Winslow, located in the recently designated Pinelands Protection Area, an ecologically sensitive multicounty region similar in purpose to the Meadowlands area. While Camden County officials had approved the Winslow site for a new landfill, the Pinelands commission eventually rejected the application because it conflicted with the regulations of the Pinelands Comprehensive Management Plan.[55] The county next entered negotiations with disposal facilities around eastern Pennsylvania, at least until development on the remaining resource recovery projects could be resumed, effectively halting development of new landfills in the county while exporting wastes outside county borders.

But here—and Camden County was not alone in this approach—an interesting strategy emerged. In comparing the outcomes of the 326 amendments' planning process, several counties implemented "interim" solutions for waste disposal that reduced the urgency for siting new disposal facilities. As Camden County found with the Kinsley Landfill, they *could* send waste elsewhere, and so they did. It is not clear from the historical record whether Camden County officials selected the Winslow site with the knowledge that it may have been undevelopable as a landfill due to the new environmental protections. But even designating a site met some of the 326 amendments' requirements, and having a site eventually disqualified on ecological grounds cleared a path for "emergency" or interim waste transfers to out-of-county locations, for a potentially indeterminate amount of time. That such a course of events might be intentional is what likely invited the NJDEP's legal and administrative action.

Other groups of New Jersey counties carried this attitude toward its logical conclusion: they did not need to develop waste disposal sites within county boundaries if they could find someone else to take it. For instance, the formation of the Hackensack Meadowlands area included provisions that virtually all municipalities in Bergen, Essex, Hudson, and Passaic Counties and a few towns in Union County be able to send their wastes to disposal facilities run by the Hackensack Meadowlands Development Commission for an indeterminate amount of time. Bergen and Essex Counties planned for a possible end to this relationship by pursuing development of waste-to-energy incinerators, commencing operations around the end of the 1980s. However, Union and Passaic Counties indicated that they intended to pursue such facilities even though little planning had yet taken place by 1985, while Hudson County, according to the NJDEP,

> was scheduled to have a resource recovery facility on line by January 1, 1985. This operational date was not met and a site for the facility has yet to be selected. No interdistrict agreement exists between Hudson County and the Hackensack Meadowlands Development Commission. Sites for the proposed resource recovery facility and ... [an] emergency backup landfill has not been designated. Hudson County is behind schedule in the implementation of its plan ... In the meantime, Hudson County must either rely on limited disposal capacity within the Hackensack Meadowlands District or reach an interdistrict agreement for additional disposal capacity in another district.
> ... Hudson County is one of [the] counties sued by the state for failure to implement their approved district solid waste management plan. Hudson County must move forward on the siting and implementation of a landfill and resource recovery facility in order to avert a potential crisis situation.[56]

By the same token there were no clear penalties for Hudson County if it *did* end up relying on disposal capacity in the Meadowlands or elsewhere, so long as the county could pay the disposal fee and the other party was willing to take the trash.

Counties in north and northwestern New Jersey adopted a similar posture. Sussex, Morris, Mercer, Warren, and Hunterdon counties all relied on exporting their wastes to landfills in Pennsylvania, or the rapidly expanding Edgeboro Landfill in Middlesex County, near the geographic center of New Jersey.[57] For Warren and Hunterdon Counties, this was in fact the product of choices made by private waste collection and hauling firms. While the counties had planned to jointly develop and operate a landfill in the town of High Point, the project fell through, prompting the NJDEP to issue an "emergency redirection order" mandating the counties' waste be disposed at facilities in Ocean County, more than fifty miles away. Private haulers rejected that option and instead chose less expensive disposal sites in eastern Pennsylvania.[58] Like some of the counties utilizing landfills in the Meadowlands area, northern and northwestern counties of New Jersey varied in their attitudes toward exporting waste rather than planning to dispose of it within their own borders. Warren and Hunterdon Counties would eventually work together to site a waste-to-energy incinerator, while Sussex County, like Burlington County, would develop a comprehensive public landfill and recycling complex. However, Morris County, like Hudson, would essentially refuse to consider development of disposal facilities within their own borders, precipitating years of conflict with the NJDEP.

In parallel with the 326 amendments' planning process, the number of landfills in New Jersey continued to shrink. The NJDEP continued to pursue vigorous enforcement of environmental protection rules, causing some landfills to close their gates. Others, typically older municipal landfills, reached their maximum capacity and either closed voluntarily or did not have their operating permits renewed. Still others were ordered closed by various components of the New Jersey court system on environmental or operating permit grounds. As described in the opening pages of this chapter, between 1970 and 1985 the number of landfill sites in the state dropped from around 370 to just 59.[59] But the reality was even more drastic than the sheer drop in number of facilities suggested: 85 percent of the now 10 million tons of waste disposed in New Jersey went to just eleven landfills.[60]

The 326 amendments themselves, along with subsequent implementation actions by the NJDEP, make clear that the state intended for each county to have a credible plan for collecting and disposing of its own waste, offering the first iterations of what would later be known as the "self-sufficiency doctrine" in New Jersey. Permanent reliance on sending wastes to another county would not be acceptable unless counties had jointly and proactively planned for such a strategy. A permanent reliance on sending wastes out of state was also considered unacceptable because those "foreign" recipients of New Jersey's trash in Pennsylvania, New York, or elsewhere were under no obligation—despite U.S. Supreme Court rulings to the contrary—to continue doing so for the indefinite future. Furthermore, out-of-state facilities had no involvement with the New Jersey waste planning process. In many ways, NJDEP's self-sufficiency doctrine represents an admirable adherence to principles of environmental responsibility: if you made the trash, you ought to find somewhere to safely dispose of it in your own backyard. At the same time, it was already clear by 1985 that developing new disposal facilities in each of New Jersey's counties might well be impossible, and that intentionally planning for out-of-county, and even out-of-state, disposal would be the most realistic option for some counties. Intentional, transparent plans to export waste would at least be preferable to the profusion of "interim" directives—some surely employed out of genuine emergency due to shrinking disposal capacity, but others adopted as a strategy to duck the planning process. In any event, the state and counties of New

Jersey would need to develop strategies for the orderly collection and distribution of wastes to disposal facilities in order to maximize the effective lives of the few that remained.

Directing Flows of Waste, 1: Flow Control

Considering the amount of landfill capacity in New Jersey was decreasing year by year, it was sensible that the state and counties developed strategies to direct the flow of waste materials to particular sites. If waste haulers could be required to dispose of the materials they collected at designated locations, rather than simply choosing the cheapest, nearest, or for whatever reason most convenient landfill, then the overall pressure on any given landfill could be reduced and the working life of that site extended by a few more months or years. This was the logic behind the state's efforts to ban disposal of wastes collected in New York, Pennsylvania, or anywhere else outside New Jersey's borders.

But waste "flow control" could also play an important role in the financing of new disposal facilities. By ensuring a captive customer base, counties (or their contractors) could be more confident of recovering their costs, and potentially even profiting, from the significant risks and investments of siting, designing, building, and operating a new landfill, incinerator, or other disposal facility.

In the era of town dumps—also a time when many neighborhoods and even entire towns did not have organized, regular trash collection—waste flow to disposal sites was mostly a function of geographic proximity or cost. The advent of state planning however introduced new thinking about waste collection and routing: if plans addressed the final destinations of New Jersey's trash, why not also seek to optimize collection and hauling routes? The 1970 SWMA extended powers to the NJDEP to regulate disposal facilities. SWMA's companion legislation, the SWUCA granted new authority to the Board of Public Utilities Commissioners (PUC; later, Board of Public Utilities or BPU) to regulate the waste management industry like a utility, as discussed above. Central to the legislation was not only authorizing the PUC to set rates for waste collection and disposal but also to designate collection and disposal "franchise areas" in which waste management companies could operate. Part and parcel of this franchising authority were powers to control exactly who was allowed to work in the industry through certifying (and revoking certification of) businesses as well as individuals and assess penalties for firms and individuals found to be operating outside of the certification rules. Finally, SWUCA granted the PUC authority to "ensure that no solid waste collection or disposal utility is permitted to limit bidding, withdraw from a specific territory, or endeavor to eliminate competition"—powers clearly intended to reduce or eliminate the murky municipal contracting practices that had been uncovered during the solid waste hearings of the late 1950s and 1960s.[61] In contrast to many regulations which exempted public utilities from conventional bidding processes, SWUCA insisted that certified waste management companies—and New Jersey's towns—engage in a public, stepwise bidding and review process.[62]

Despite the potential benefits of state or county flow control, it was not entirely clear whether SWUCA or the 326 amendments planning process legally permitted flow control. In 1980 the NJDEP commissioner Jerry F. English wrote to the office of the attorney general, requesting a formal opinion "interpreting the Solid Waste Management Act and the Solid Waste Utility Control Act to determine whether solid waste management districts, acting pursuant to solid waste management planning, have the authority to require that solid waste generated within the districts be directed to specific waste disposal facilities."[63] Given that these acts and the subsequent 326 amendments were understood as "mandating a regional planning approach as a basis for solid waste collection and disposal throughout the State," Attorney General Degnan opined that "the clear objective is thus to commence formulation of a management plan which most effectively and economically controls waste collection and disposal."[64] Furthermore,

> As an integral part of this planning process, the district is to develop a strategy to most effectively provide waste disposal services to the region . . . The management plan developed by the district may therefore provide for the channelization of wastes to specific facilities if such planning is reasonably deemed to best effectuate the regional strategy so formulated . . . As a result of district planning, a waste management strategy directing the solid waste stream to specific facilities may be developed by the districts . . .
>
> It is therefore our opinion that the Solid Waste Management Act and the Solid Waste Utility Control Act establish the authority of the solid waste districts through their comprehensive planning to direct the flow of wastes to selected destinations.[65]

Six months later in June of 1980, Commissioner English requested another opinion from the office of the attorney general, as to whether the SWMA and SWUCA authorized the BPU to establish "uniform average rates" for waste disposal within a single management district, "even though independent rates are set for each solid waste facility."[66] Because New Jersey had chosen to regulate waste collection and disposal as utilities, companies operating in this area had their rates set by the BPU. But in contrast to conventional utility regulations, which focus on monolithic entities like electricity or natural gas providers typically operating as monopolies with designated, noncompetitive service areas, several waste collection and disposal companies might exist within the same county or even the same town and compete with one another on price and type of service. At question was whether the BPU had the ability to assign "uniform rates" for an entire county while still reflecting that individual waste management companies would have distinct rate structures based on their costs, equipment, operating practices, and so forth. Attorney General Degnan wrote that

> methodologies can be devised to pass on to consumers a uniform cost of service even though each facility operates pursuant to an independent rate schedule . . . If proposed by a district, and approved by the DEP, this "weighted average" approach would calculate an equalized charged to be paid by consumers, with all such revenues distributed by an implementing agency to facilities within a district based upon a formula encompassing such variables as wastes received over a specific period of time and the independent rate base of each facility. Similarly, the BPU through its franchising powers may equalize or control costs within a region by directing wastes to specific facilities . . . and too, uniform rates may also be set if the solid waste facilities are public authorities.[67]

The logic behind this position was that a uniform or equalized rate could mask some of the implications of flow control as well as the risk inherent to developing new waste disposal facilities. If a facility was overburdened and state and county officials thought it prudent to direct waste to another location, the overburdened facility would not necessarily miss out on revenue. Alternatively, directing wastes to a new, higher cost disposal facility would, under a *normal* rate structure, cost consumers more. But with uniform rates, facility users— and ultimately, taxpayers and businesses in New Jersey—would pay the same for waste disposal within a single county regardless of whether the waste went to a high-tech waste-to-energy incinerator or an old landfill at the end of its working life. In the same fashion, uniform rates would subsidize the operations of new waste management facilities as they commenced operations, and old ones when they started to wind down at the end of their working lives, both of these being periods of time when the amount of waste disposed at the facility—and thus facility revenues—might be limited. Under a *normal* rate structure, facilities receiving less waste would receive less revenue as well, a reality that could complicate the financing of new disposal facilities and make them less attractive to investors. Conversely, uniform rates could subsidize poorly managed facilities and reduce the incentive for innovative, more efficient new disposal technologies.

Perhaps unsurprisingly, the issues of flow control and uniform rates attracted immediate legal challenges. One of the first suits was brought by a group of waste collection firms and landfill operators in conjunction with Union County and towns within Union County including Elizabeth and Kenilworth. In this case, known by its shortened title of *Mastrangelo vs. NJDEP*, the appellants alleged that the SWMA and SWUCA as amended did not afford the NJDEP authority to require individual waste collection firms or county governments to direct flows of waste to particular disposal sites.[68] On submitting their initial "326 plans," the NJDEP found that Union County and Middlesex County (host to the Edgeboro Landfill, one of the state's largest) needed to develop an agreement with each other for Union County to send approximately half of its waste to Middlesex, for at least a twenty-year period. Union and Middlesex Counties failed to adopt an agreement on their own accord, and so NJDEP imposed their desired arrangement on the counties and waste management firms operating in those counties.

Union County and the other appellants claimed that the imposition of flow control contradicted the 326 amendments' intention of county-led solid waste planning. The Supreme Court of New Jersey, however, disagreed. Justice Robert L. Clifford wrote in the unanimous opinion that

> The principal thrust of the [flow control agreement] is the redirection of solid waste flows away from the Hackensack Meadowlands District (HMD), an approach that appellants view as untenable. The [NJDEP] cites several factors in support of its position. Initially, DEP notes that the Hackensack Meadowlands Reclamation and Development Act requires that HMD develop waste disposal facilities sufficient to enable it to continue to receive waste being deposited in the HMD by other districts . . . In light of this the DEP contends that an inclusion of additional solid waste streams not statutorily mandated would reduce the remaining useful life of available HMD landfill capacity to a dangerously low level. DEP also maintains that a policy allowing Union to continue its dumping in the HMD would be tantamount to an endorsement of an existing attitude among districts that envisions the Hackensack Meadowlands region as the dumping ground for New Jersey. This in turn would reduce the incentive of other districts to become more efficient in their disposal strategies and to develop alternative solutions to solid waste disposal.[69]

While upholding the flow control arrangement, the court simultaneously struck down the NJDEP's ability to independently develop and impose such arrangements. The logic behind this curious decision was that the SWMA and SWUCA laws actually afforded the BPU that power, not the NJDEP.

> These circumstances lead us to conclude that DEP has the authority under the Solid Waste Management Act to provide general direction with respect to interdistrict waste flow. By "general direction" we mean the power to assess the solid waste disposal capacities of each of the solid waste management districts and to direct any district to dispose of its waste in any other district. That power can be said reasonably to flow from the Act.

On the other hand, the authority to direct individual solid waste collectors to collect and transport waste streams to specific disposal sites and require or designate specific disposal facilities as the ultimate destination of particular waste streams is more appropriately within the realm of BPU, given its responsibility to weigh and regulate the economic aspects of the solid waste industry. If specific resource recovery plants or other collectors or disposal facilities believe that they need a franchise grant to assure economic survival, they should apply to the BPU on a case-by-case basis.[70]

Accordingly, the court insisted that the NJDEP and BPU promulgate essentially joint guidelines for crafting flow control agreements by the end of 1982, which the agencies accomplished in short order.[71] In any event, as the number of disposal facilities in each county diminished further during the 1980s, typically a single facility in each county was designated the de facto destination for all wastes collected in that county.

By 1985, flow control, along with the self-sufficiency doctrine, was well established as a cornerstone concept in managing New Jersey's trash. While self-sufficiency represented the guiding principle for the state's system, flow control was the primary tactic through which county-based planning would be implemented, financed, and enforced. Upping the ante even further, the NJDEP's 1985 Solid Waste Management Plan Update declared that flow control would not be used to enforce permanent regionalization of the waste management system. That is to say, "Those districts which have developed and are implementing plans will not be penalized for the lack of action on the part of other districts"[72] and be forced to take the trash of others over the long term. Furthermore, the agency declared that counties facing disposal facility closure in the foreseeable future should not assume that the NJDEP would approve plans to simply export waste to another county or out of state. Newly confident from the *Mastrangelo* supreme court decision, the 1985 plan update went on to say

The Department [of Environmental Protection] and Board [of Public Utilities] have, in a number of instances, used the waste flow direction power to redirect wastes from landfills which were closing. However, the Department has always taken the position that the responsibility for determining the site for disposal of a district's waste remains with the district. It has been the policy of the Department that each district should be monitoring the conditions and remaining capacity at disposal facilities which serve it and determining the next available facility (as is required by law) . . .

Emergencies, such as landfill fires or other unanticipated closures can trigger short-term (180 days or less) redirections by the Department and Board. Such emergency redirections must, according to the rules, be incorporated into the relevant district plans. However, in most cases, districts have not done so.

Since provisions of disposal capacity (or arrangements therefor) is one of the most important roles delegated to solid waste management districts, this violation of state rules and directives appears to signal unwillingness on the part of some districts to assume one of the basic responsibilities delegated under the Solid Waste Management Act.[73]

The NJDEP was vigilant in upholding this precept regarding waste flow.[74] For instance, in 1984 when Camden and Salem Counties sought to use Gloucester County's (privately owned) Kinsley Landfill for disposal, the agency through the New Jersey Superior Court sought injunctions to permit this use only on the condition that Salem and Camden site disposal facilities of their own. In a curious twist, the superior court also ruled that *only* municipalities or counties having reached agreements with Gloucester County over disposing waste at Kinsley Landfill would be permitted to do so. This ruling effectively banned the import of wastes from other states including once again, the city of Philadelphia, which had refused to pursue such agreements.[75]

By the mid-1980s it held that the NJDEP, BPU, and county government entities entrusted with solid waste management planning had the legal authority to direct flows of waste, both between counties and within individual counties to particular disposal sites, pursuant to the waste management planning process. However, there was, and remains, another important dimension to state and county oversight of waste flows: illegal and uncompetitive collection practices.

Directing Flows of Waste, 2: Cleaning up Collection Practices

Devising a plan for where waste should go was itself a major undertaking by the state and New Jersey counties. However, ensuring that wastes were actually collected and disposed of accordingly would prove to be an entirely different matter. Inconsistent waste collection practices would mean that all of the planning and projections undertaken about remaining landfill capacity, out-of-state waste imports and exports, and regulating rates charged for waste services were purely academic exercises. The waste collection process in New Jersey, until the advent of a formal planning process, was haphazard and inconsistent. As hearings held in the late 1960s (discussed in chapter 2) illustrated, contracts for waste collection and disposal were frequently opaque. While it was clearly possible for collection companies ("haulers") to compete with one another for a town's business on the grounds of price or service quality, frequently town hauling contracts would receive only one bid. As early as the late 1950s, components of state government were trying to understand and explain the reasons behind unusual patterns and practices in waste collection. In a 1959 report, Deputy Attorney General John J. Bergin noted that the disorganized, highly local nature of waste management planning in New Jersey's towns and counties was vulnerable to all sorts of abuses from financial misfeasance to capture by organized crime groups. He wrote, "It is respectfully recommended that legislation be enacted stabilizing and controlling this rapidly expanding industry that so vitally affects public health and morals, keeping in mind our traditional respect for free enterprise. It is evident that many of the contractors, municipal and otherwise, prefer genuine competitive bidding. They would rather rely on their skill, experience and other legitimate factors to insure their success. Many essentially honest people in this business have had to face a clear cut choice; either go along with the conspiracy or go out of the only business they knew."[76] So-called uncompetitive behavior arose most frequently from agreements between waste haulers to divide towns or regions of the state into territories, wherein one member of the cartel would be the sole bidder for any contract, or at least the sole credible bidder for any contract. As early as 1965, trials around price-fixing and hauler collusion were unfolding. In one prominent case centered on haulers in greater Philadelphia but disposing wastes in New Jersey, the U.S. District Court for Eastern Pennsylvania upheld an interpretation of the Sherman Anti-Trust Act applied earlier by prosecutors in lower courts.[77] In that instance, a group of haulers acting in concert as the Pennsylvania Refuse Removal Association was sued by the U.S. federal government on the grounds that "that members of the Association, including the defendants, agreed to fix prices, allocate customers and rig bids in the refuse removal business in the Greater Philadelphia area, practices proscribed by §1 of the Sherman Anti-Trust Act."[78] Per the opinion of deciding Judge Alfred L. Luongo,

> As to price fixing . . . [Witness] testified that at an Association meeting in December 1962, defendant Graziano recommended to the membership that prices for refuse removal be raised by between 25% and 50%, effective January 1963. According to [witness], the membership, including all the defendants, expressed agreement by raising their hands . . .
>
> In this same regard another witness . . . testified that at a meeting in his office in August or September 1962, attended by defendants Graf, Graziano and Vile, Graziano, when asked what the purpose of the Association was, stated that it was "being formed to fix prices in our industry . . ."
>
> As for allocation of customers . . . an Association member testified that at a membership meeting in 1962, defendant Graziano proposed that one member should not take a customer of another member and that if he did he would either have to give the stop back or make amends either by a money payment or by giving up equivalent work. [Witness] further testified that defendants Graf, Graziano, Vile and Coren all signified their agreement with this proposal by raising their hands . . .

Similarly ... a Pennsylvania refuse remover, testified that he went to a meeting of the Association because he was losing accounts in an area known as Hartsville Park. Defendant Graf assured him that if [witness] joined the Association he would lose no more business. [Witness] joined but continued to lose accounts, especially to one Donald Shindlar, an Association member. [Witness] informed Graf of his continuing misfortune whereupon Graf arranged to meet with [witness], Shindlar, and Shindlar's driver. The meeting took place in Graf's car in Hartsville Park. The four men drove around and [witness] pointed out to them the stops he had lost to Shindlar, both before and after becoming a member of the Association. Graf then instructed Shindlar to give back to [witness] those stops taken from him after he became a member, but that he could keep those he had acquired before [witness] joined the Association ...

In the area of rigging bids, there was testimony by [witness] that at a meeting in 1962, defendant Graziano told the membership "that if a brother member was hauling a stop and three or four other members had received the bid, that they should get in touch with the member that is hauling the stop and find out what he is bidding and bid higher, so that the member could keep the stop." Present when this statement was made were the other defendants, Graf, Vile and Coren.[79]

In this instance, defendants were found guilty of violating antitrust laws. Similar legal efforts were undertaken in the 1950s and 1960s in New Jersey, but with limited success,[80] even as a great deal of investigative work into the hauling industry was completed during this time. One report filed in 1981 by the Task Force to Study Solid Waste Regulation—whose membership was comprised of representatives from the Department of Law and Public Safety, the NJDEP, and BPU— articulated a clear origin story for uncompetitive behavior in New Jersey:

The 1940s and 1950s witnessed more sophisticated equipment, the unionization of many firms and a substantial increase in the number of municipalities awarding contracts for solid waste collection especially in the northeastern part of the State ... With the increase in the number of contracts came an increase in the number of firms. This in turn provided an incentive for the older firms to band together to keep the new comers in their place. This solidarity was allegedly effected through the union for solid waste firms, Teamsters Local 945, and a new association formed in 1956 called the Municipal Contractors Association (MCA). Together and separately they purportedly used a variety of practices to maintain the 30 or so MCA members' stranglehold on the municipal contract market. These practices included denying dumping privileges to non-association members who bid on municipal contracts and tailoring bid specifications to incumbent collectors. The result of these practices were rapidly escalating prices for municipal contracts.[81]

The 1970 SWUCA and companion SWMA can be interpreted in part as the state's response to these types of collection practices, which afforded the NJDEP and BPU powers to license and revoke licenses of firms active in waste hauling. The SWUCA also afforded BPU the authority to create "franchise areas" for haulers, legally protecting a hauler's service area from competition. Franchises were intended to both incentivize investment in improving collection efficiency (by limiting the ability of a competitor to take business) and also impose some measure of state oversight of haulers' operations.[82] Nevertheless, while competition between companies for waste collection contracts within any given town was clearly possible due to the franchising provisions of the 1970 SWUCA, it remained relatively uncommon.

Additional hearings and investigations during the 1970s and 1980s sought to understand the nature and extent of uncompetitive behaviors.[83] These efforts culminated in two related legal outcomes. The first was the indictment of nearly sixty individuals by October of 1980 under New Jersey's antitrust laws. The majority of those indicted entered guilty pleas as part of agreements made with the state, acknowledging that they or their companies had indeed participated in uncompetitive behaviors of the types outlined above. During the limited numbers of trials resulting from the indictments, an additional motivating factor for uncompetitive behavior also emerged—organized crime.

For decades, the involvement of various organized crime elements had been an open secret in New Jersey's waste management economy. Holding close ties with groups in New York City and Philadelphia, at its most fundamental level the function of organized crime was creating and enforcing the uncompetitive markets for waste collection and disposal—in other words, serving the same functions of bid-rigging, price-fixing, and customer allocation as the Pennsylvania Refuse Removal Association or the Municipal Contractors Association. The key distinction around organized crime was in the frequently violent and extortionate means to these ends and the reinvestment of revenues from the waste business into other illicit activities. During the trials of Louis Mongelli, Louis Spiegel, Anthony Scioscia, and John Gentempo, who had each refused to enter into plea agreements with the state over the 1980 indictments, presiding judge Arthur S. Meredith took great pains to make clear that the defendants were not on trial for connections to the crime underworld, but rather for violating New Jersey's antitrust laws. However, it became clear that at times the two were inseparable. From news coverage of the trial,

New Jersey's trash wars, in which three refuse company owners have been slain gangland style in recent years, have repeatedly been linked by state prosecutors and in Congressional testimony to members of organized crime. But the judge forbade any mention of the underworld before the jury during the six-week trial, which was held in Somerville, to protect the rights of the four defendants . . . Accordingly, before the trial began, Judge Meredith granted defense motions to suppress as prejudicial testimony before the jury mentioning organized crime.

But gangster influences in the trash industry emerged at the trial inadvertently. On the witness stand, Clemente Pizzi, a trash hauler appearing as a prosecution witness, was asked why he had joined the waste association. "There had been a murder," Mr. Pizzi started to reply, suggesting that his reason had been fear. But he was abruptly cut off by Judge Meredith. The reference apparently was to the unsolved gangland-style murder in 1976 of Alfred DeNardi, a New Jersey carting company owner who prosecutors said had broken the trash industry's rule against bidding for portions of another waste hauler's route. The state also is still investigating the deaths of two other New Jersey garbage collectors, Gabriel San Felice in 1978 and Crescent Roselle in 1980, who were slain under similar circumstances.

Mr. Pizzi's comment, like the more oblique testimonial references of other witnesses to underworld violence, brought a flurry of defense motions for a mistrial. Since the trial began Feb. 7, there were more than 50 motions for mistrial, all denied.[84]

It is within this context of uncompetitive behavior that the participation of organized crime elements in the waste management industry must be understood. Certainly, stories about organized crime in waste management are salacious and colorful;[85] equally certain is that there were real murders, violence, extortion, and other types of harm perpetrated by gangsters on one another but also members of the public and representatives of various levels of government in pursuit of control of the waste industry.[86] While the impacts of individual violent acts were confined to victims and their families, it is also the case that uncompetitive behavior in waste hauling imposed additional costs on the rest of society as well, by raising prices for waste collection service and upholding inefficient collection practices and routes. There was a time when apparently no customer could be considered immune from extortionate practices. Relating a story from the late 1970s, the New Jersey State Commission of Investigation reported that

> The Children's Specialized Hospital of Mountainside became dissatisfied with its collector, Statewide Environmental Contractors, a member of the [New Jersey Trade Waste Association, TWA, a price-fixing cartel]. In the course of a year, the hospital, which had been paying $400 per month for collection services, found an alternative collector, a member of the Hudson County

Association [HCSA, a competing cartel] that was willing to service the hospital for $600 per month. The first collector requested a grievance proceeding which was held before members of the TWA and HCSA. Based on "property rights", the second collector was ordered to give the "stop" back to the previous carter. Unsuccessful in a year-long search for a third hauler, the hospital now found itself compelled to utilize the services of the original contractor—only now it was required to pay $800 per month, double what it had been paying Statewide Environmental Contractors before it sought another carter.[87]

Stories like these were not uncommon, and paired with the outburst of violence in the late 1970s and early 1980s, a second major outcome of the investigations into uncompetitive waste hauling in New Jersey was launched, aimed at rooting out organized crime. The Waste Industry Disclosure Law of 1983, better known by its legislative coding as the A-901 Act was the first of its kind in the United States. Introduced by Assemblyman Raymond Lesniak, A-901 added to the licensing requirements imposed by the 1970 SWMA and enforced by the NJDEP and BPU, by requiring participants in the waste management industry to attest that they were free from "criminal records, habits, and associations" linked to organized crime.[88] Per the new rules, "the procedure requires that the companies, and over 6,520 of their officers, directors, partners, key employees and holders of equity or debt liability submit disclosure statements to the NJDEP. The Attorney General, through the Solid/Hazardous Waste Background Investigation Unity of the Division of State Police, performs a background investigation including a criminal records check. The Division of Law in the Attorney General's Department of Law and Public Safety evaluates the information revealed by each investigation and prepares a report in which is advises whether NJDEP is precluded by the standards of A-901 from granting a license."[89] After surviving a legal challenge, A-901 rules became law in 1986, and state officials immediately took action. By 2011, twenty-five years after the implementation of the law, approximately 175 persons were banned, had licenses revoked (or denied on application), or otherwise debarred from the New Jersey waste industry. Some sixty others either withdrew voluntarily from the licensure process or were granted probationary licenses "pending remediation of potentially disqualifying factors."[90]

Despite this success, seeking to clamp down on the activities of organized crime is akin to slapping a pool of water—while there is a clear and sometimes impressive immediate effect, the displaced water simply finds its way to new areas. The A-901 process was immediately recognized as well-intentioned but complex, and rife with loopholes. Authority was divided between three different components of state government, and backlogs and delays quickly became overwhelming, at times limiting the impact of the law.[91] As the waste industry started to understand the impacts of the law, complex new ownership and leasing arrangements emerged that evaded both the letter and the spirit of the new rules. For instance, the New Jersey State Commission of Investigation reported in 2011 that

> Joseph Virzi and Henry Tamily are convicted felons debarred from the solid waste industry in New Jersey, but they, and their business partner, Marino Santo, continue to profit from it as principals in a realty concern, which receives substantial rent from a solid waste and recycling enterprise that acquired their former trash-hauling businesses. Owners of this multi-firm enterprise include their adult children. The Commission's inquiry also revealed evidence that the younger family members in this circumstance have engaged in the improper sharing of official DEP vehicle credentials. Known as "decal fronting," this practice enables unlicensed and out-of-state firms to dump solid waste in New Jersey landfills while avoiding both A-901 scrutiny and the payment of disposal fees assessed against legitimately credentialed haulers.
>
> Once based in Manhattan where they prospered for decades under organized crime's property-rights system, Tamily and Virzi pled guilty in 1997 to charges of attempted enterprise corruption stemming from the first major investigation of mob control of New York City's garbage-hauling industry. They each received indeterminate prison sentences of 18 to 54 months and consented to lifetime debarments by the city. Santo, their business partner for nearly five decades, escaped indictment but nonetheless was also debarred by the city for life. In 1999, solid waste regulators in New Jersey debarred each of the men for terms of five years. None has sought to return to employment in the industry here. New York authorities took the additional step of requiring Santo, Tamily and Virzi to sell their waste-hauling businesses. The buyer was Jem Sanitation Corp. of Lyndhurst, N.J., which paid $900,000 in the combined purchase. Jem's parent company—Jem Carting Group Corp.—is owned and managed by a group that includes the adult children of Santo, Tamily and Virzi. It consists of various entities, including paper recyclers and solid waste haulers, tied to a common address—an industrial complex at Schuyler and Page Avenues in Lyndhurst. They pay combined rentals of $120,000 annually to a landlord, Schuyler-Page Realty Co., which is owned by the three debarred haulers/parents. The property in question is the firm's only real estate holding.
>
> In the course of examining this matter, investigators also discovered evidence that official truck decals explicitly issued by DEP to one licensed solid waste company in the Jem group—Jem Sanitation of New Jersey—have been provided for use by a cluster of other trash hauling firms under questionable circumstances. Those companies—Classic Recycling of New York Corp., Classic Demolition Co. Inc. and Classic Sanitation Co. Ltd.—are owned or operated by relatives of Marino Santo. Although these firms have

administrative offices in Clifton, N.J., and park some of their vehicles in New Jersey, none is licensed to haul and dispose of solid waste in New Jersey. All of their customers are in New York City. The decal-sharing arrangement ostensibly was effectuated through truck leasing agreements established in 2007 between Jem and the trio of Classic companies. But the Commission found that these agreements, though signed and seemingly official, were bogus documents used as a cover for enabling the Classic companies to bring solid waste from New York City into New Jersey and to dump it at landfills and incineration facilities here without a license, without A-901 integrity vetting and without having paid New Jersey licensing fees.[92]

Beyond the challenges of enforcement, there were significant flaws with the A-901 rules themselves, as elements of organized crime that had initially been active in hauling were largely pushed out of that sector only to resurface—legally—in the worlds of recycling and management of construction & demolition debris.[93] Until approximately 2016, A-901 rules did not apply to firms dealing with recyclables, which the legislature had defined as a separate set of materials in the 326 amendments.[94] There was also a loophole centered on fill dirt, which nefarious characters were exploiting to remove contaminated soil from toxic cleanup sites for disposal in New Jersey landfills, resale as "clean fill" for construction projects, or more commonly, dumping in abandoned lots and sensitive ecological sites.[95] The environmental challenges of "dirty dirt" became especially salient in the waste management aftermath of 2012's Superstorm Sandy.

In any event, it is inaccurate to characterize all of the uncompetitive and illicit behavior as having links to organized crime. But the specter of violence and extortion has surely left an indelible mark on the waste management industry in New Jersey and the public's perception of the industry. Former deputy attorney general Stephen Resnick testified to the State Commission of Investigation in 1989:

> Yes, I definitely see [organized crime] involvement in the industry, and let me just say that at different times and in different parts of the State it can be more pronounced. In other words, it is not a total complete control of the State, nor even a total complete control of any part of the State. It manifests itself in different locations at different times . . . We [the Division of Criminal Justice] become aware of an allegation. We pursue it. In some of these matters that allegation has led us to evidence of the more traditional organized crime involvement. In others it has not. We don't know if it's there; we don't have any evidence that it's there. Yet in other cases we found an aura of that involvement, but when you investigate intensively, you don't find the substance.

And it's entirely possible in one of these cases that a fellow was just making use of the organized crime impression to get his way. We just don't know. But the fellow who did this . . . shows up at a garbage company in a limousine; two fellows get of the car before him and stand at the door; he marches in, talks tough, and we come to find there's no substance behind it. None. Not even a shred . . .

The influence is there, but it does not permeate and control the entire industry . . . The bulk of the companies in this State engaged in this business are small operations that mind their own business and are not mob-tied.[96]

"Mob-tied" or not, we can observe many ways in which haulers ignored and continue to ignore state and county rules for collection and disposal in ways that undermine the intentions of the state's waste management planning process.[97] These impacts range from violating flow control orders and illicit dumping to economic frauds like fixing scales, mixing loads of trash and recyclables, or illegally exporting waste out of state. Per the State Commission of Investigation's 1989 study,

> there is a tremendous economic incentive for haulers to violate waste flow
> orders. And given the woefully inadequate size of the BPU and DEP enforce-
> ment staffs, there is little fear of discovery and sanctions. Many collectors
> are continuing to bill their customers the full cost . . . even though they are
> depositing the trash at unauthorized, but cheaper disposal facilities . . . the
> ability to successfully avoid the waste flow order becomes all-important.
>
> State officials have publicly stated that, as of March 1988, up to 20 percent of
> the waste that should have been going to county transfer stations was being
> shipped directly out of state in contravention of waste flow orders . . . In
> May 1988 officials of the Hackensack Meadowlands Development Commission
> (HMDC) announced that trash volume . . . had increased by almost 50 percent
> since March 1 1988. This astronomical swell in trash flow to the relatively
> inexpensive HMDC landfill . . . coincided with Bergen County's opening of a
> nearby transfer station where the dumping rate is nearly four times higher.
> Meanwhile, the Bergen facility's trash flow immediately dropped almost
> 40 percent.[98]

In these instances, several commentators have observed that economic frauds common to the New Jersey waste management industry are symptomatic of fundamental flaws with the regulatory scheme itself. In the past, even BPU staff themselves noted that rate setting by their agency was too slow to adjust to changing prices in the disposal marketplace. For example, in 1988 Fred S. Grygiel, the chief economist for the BPU, observed:

[We] shouldn't be too surprised if people take their waste to the least cost option . . . that's what we should expect, notwithstanding a piece of paper that says you are not supposed to do that. The motivation is obvious. They would like to increase the margin at which they are operating. They can do that simply by violating an order. The order, on its face, may be unsound economically; it may attempt to be doing something that is so offensive to rational economics that even economists would recommend that people violate the law and do it, because society would be better off if they did . . .

The . . . use of waste flow orders is, to me, just not a viable way to go. Attempting to enforce a waste flow order in an open economy . . . would require a legion of people checking trucks and going through garbage to find out where it came from.[99]

Together, the 1970 SWMA and SWUCA along with the 326 amendments' planning process and provisions for flow control relocated much of the conceptual and legal power over the waste management industry into state entities like the NJDEP. Even as county governments were held responsible for planning for their own waste management futures, this was undertaken in the knowledge that each county's plan would reflect the overarching goals and desires of state officials, and that these goals and desires could be upheld in a court of law. Yet, similar levels of control could never be brought to bear on the hundreds of individual waste hauling firms in the state. The inability to force haulers to comply with waste management plans and directives—which, at times, ran counter to economic common sense—would pose significant challenges to the planning process during the 1990s and 2000s.

In any event, efforts at state- and county-level planning sought to impose some order on the process while simultaneously moving to shut disposal sites understood to be hazardous for environmental and human health reasons. But even with all the planning, flow control, attempts to limit out-of-state wastes, efforts to eradicate organized (and disorganized) waste hauling crime, and extensive legal cajoling that accompanied each of these action areas, the waste "crisis" in New Jersey was far from resolved. The 1985 Solid Waste Management Plan Update estimated that during the period 1985–2000, New Jersey would experience a "shortfall" of disposal capacity of nearly 50 million tons, even including all planned new disposal projects.[100] That is to say, there would be 50 million more tons of trash needing disposal than there would be available places to put it. Very limited success in siting new sanitary landfills was perhaps anticipated by policymakers in this most densely populated state. While sanitary landfills represented a marked improvement over the town dump, if it was near-impossible to see them built, could they really represent a long-term solution for New Jersey? As Deputy Attorney General John J. Bergin predicted in his report from 1959, "With the millions of tons of refuse collected daily the time is bound to arrive when there will be no dumping space left in the small State of New Jersey. Construction on this type of base is generally limited to parkland, air fields, parking lots and stadia. Consequently a different method of disposal will have to be looked toward."[101] The challenges of developing large regional landfills forced policymakers, private firms, and members of an increasingly alarmed public alike to examine the possibilities of recycling and incineration in much greater detail.

4

Recycle or Incinerate?

● ●

> For years we have run away from the
> hard choices required for a solution . . .
> In just a few years, the issue will no
> longer be which county takes which
> garbage; instead, it will be where to find
> room to put <u>any</u> of the garbage—even if
> we all agree to share the burden equally.
> We can no longer walk away from the
> problem; it has arrived on our doorstep.
>
> Nobody likes garbage. Everyone creates
> it. Everyone must cope with it.
> —Governor Thomas H. Kean,
> Annual Message to the New Jersey
> State Legislature, January 10, 1984

These words from Governor Thomas H. Kean in 1984 recognized the realities of the garbage crisis in New Jersey. Despite intensive efforts to plan for new landfills, stem the arrival of out-of-state wastes, and improve the efficiency of waste collection (while rooting out elements of organized crime), the lack of disposal capacity had indeed brought the garbage crisis to everyone's doorstep. Governor Kean's address the following year also devoted considerable attention to the issue:

Last year, I told you that garbage was at the point of crisis. In the last 12 months, this has become self-evident. As a state, we have begun to learn, as our neighbors in New York and Philadelphia are learning, that garbage does not just go away. It must be put somewhere and that somewhere must be safe. Too often in the past, the poorly-sited landfills in New Jersey and other states have turned into Superfund sites. No more. In the last three years, my administration has closed almost 100 landfills. The reason? They are a danger to the public—either because they have exceeded capacity or because they endanger drinking water. In the past year, 95 percent of New Jersey's garbage was dumped at 13 sites. Six of those have reached capacity.

New Jersey can no longer avoid two central facts about solid waste. First, it will be more expensive to dispose of garbage in this and every other state. Second, meaningful action will require courage by elected officials at <u>all</u> levels of government.[1]

The limited number of suitable sites for new landfills, paired with the two-headed political monster of reluctant county governments and strong public opposition, all but demanded alternative options for disposal. The governor identified the two most promising landfill alternatives for New Jersey: recycling and "resource recovery," or waste-to-energy (WTE) incineration. This chapter examines the histories of recycling and WTE in New Jersey, major portions of which unfolded in parallel with the planning and flow control measures discussed in chapter 3. Two outcomes of these histories are important to note at the outset. First, New Jersey passed in 1987 among the first mandatory recycling laws in the United States. Second, the state planned for each county as well as the Hackensack Meadowlands Area to host a WTE facility, yet only five ended up being built by the early 1990s. As this chapter makes clear, these outcomes did little to *solve* the waste problem in the state, as the two landfilling alternatives were made to compete with each other in the public's perception and also from financial, operational, and environmental perspectives.

These outcomes are highlighted here because it is my contention that while recycling and WTE were, and are, portrayed as motivated primarily by concerns over the environmental performance of landfilling, the reality is that environmental aspects of these technologies were, and remain, a secondary consideration in the minds (if not words) of most officials tasked with addressing the solid waste crisis. That is not to say that officials, along with private firms and members of the public, were insincere in their hope that recycling and WTE could represent improved environmental performance over dumping and landfilling. Rather, it is to say that both recycling and WTE are best understood in the context of solving the disposal capacity problem, even as environmental concepts were frequently mobilized to both support and oppose these landfill alternatives. This is an important point to keep in mind, as these landfill alternatives were frequently evaluated by the public along the lines of their environmental characteristics. As this chapter illustrates, within policy-making circles, recycling and WTE were understood first and foremost as safety valves for the landfill capacity crisis. But tension and uncertainty about the purpose of landfill alternatives has come to permeate debate and analysis of New Jersey's waste infrastructure since the 1980s. Is the goal of new investment into the state's disposal capacity to reduce long-term costs? Generate electricity and steam? Conserve raw materials and natural resources imported from elsewhere? Support slowing sectors of the economy? Or simply reduce reliance on in-state landfill sites?

We can locate evidence of this confusion regarding the purpose of landfill alternatives even in the foundational documents of waste management planning in New Jersey. The 1972 Musto Commission report decries the fact that "not more than 10 percent of New Jersey's domestic solid waste volume is treated by the three municipal and 6,400 small private operating incinerators, or is subject to recycling through the 114 public and voluntary organized recycling programs."[2] While noting that various resources comprising the roughly 430 million tons of New Jersey waste to be disposed between 1970 and 2000 "are estimated to be worth over 6.4 billion dollars," the Musto Commission saw the ultimate value of recycling and incineration as "processing operations with the capability of reducing the volume of waste requiring ultimate disposal and thus conserving landfill acreage."[3]

Shortly thereafter, the 1975 Solid Waste Management Act Amendments (the "326 amendments" examined in great detail in the previous chapter) defined the terms "resource recovery" and "recycling facility" as specific components of the waste management arsenal, going so far as to retitle the 1970 Solid Waste Management Act to include the words "resource recovery."[4] The 326 amendments included in the preamble to the legislation that "it is the policy of this State to . . . Encourage resource recovery through the development of systems to collect, separate, recycle and recover metals, glass, paper and other materials of value for reuse or for energy production."[5] Readers will recall, however, that the main thrust of the 326 amendments was to insist on county-level waste management planning, and in particular planning for maximizing landfills' working lives. The 326 amendments highlight a strategy that would be taken up by many different stakeholders, casting landfill alternatives in supporting roles for other sectors of the state's economy. These technologies and processes were framed as supporting manufacturing in particular, but also industries as diverse as construction, energy generation, and chemical or pharmaceutical production.

The multiple personalities of landfill alternatives would become a regular feature of governors' speeches and proposals. For instance in a section of his

1978 "Annual Message to the Legislature" focused on expanding both fossil fuel and renewable energy production in the state, Governor Brendan Byrne pitched WTE as the technology "for converting solid waste into marketable resources and energy," before insisting that "economic incentives must be created this year to use garbage as a resource rather than a burden, consuming landfill after landfill."[6] Later on in the same speech, Byrne proposed to lawmakers that "comprehensive legislation be adopted to provide greater incentives to use garbage for energy reclamation and recycling rather than disposal in landfills."[7] A 1984 pamphlet released by the New Jersey Department of Environmental Protection (NJDEP) warned that while "the state as a whole has an average of two years of landfill capacity left," a strategy of jointly pursuing "waste-to-energy incineration of solid wastes in concert with greater reliance on source separation and recovery of recyclables" could reduce the amount of material going to landfills by 25 to 45 percent while having a "beneficial effect upon the economy of a host municipality" as a "valuable tax ratable."[8]

Despite the ominous tone of some NJDEP publications and even Governor Kean himself in the addresses highlighted at the opening of this chapter, from some perspectives the clouds were parting by the late 1980s. For instance, by 1988, Governor Kean was arguing that "last year was the year we turned the corner on the garbage problem . . . We made gains on every important front: recycling, resource recovery plants and landfills. They must be spoken of together because solving our garbage disposal problems depends on all of them. We must recycle to reduce the amount of garbage that must be burned or buried. And even with our state-of-the-art resource recovery plants, we will still need a place to put what cannot be burned, or what remains after we burn . . . Now New Jersey will be able to reduce the amount of garbage it must throw out because it is increasing the amount of garbage that it will reuse."[9] Governor Kean painted a picture of recycling and WTE working together with landfills to divert the disposal capacity crisis while also encouraging "resource recovery" and economic activity. Central to Kean's vision was a plan for each of New Jersey's twenty-one counties along with the Hackensack Meadowlands area to host a new WTE project. Not all were convinced, however, and elections in the late 1980s saw the waste management issue become politicized at the state's highest levels. Waste management appeared as a distinct plank in both Republican and Democrat platforms. The 1989 Democratic Party platform, for instance, refuted the *gains* that Governor Kean claimed:

> New Jersey is facing a solid waste crisis . . . Republican leadership has abjectly failed to clean up the solid waste hauling industry and has left county governments to struggle with a problem that demands statewide coordination and planning. This perfect example of a failure to govern must be redressed by a Democratic administration. The only thing that the State has done is to

encourage a headlong rush to build an incinerator in every county—a process that will cost several billion dollars . . . This madcap, expensive directive by Republican leadership must be halted with a regional environmentally-conscious plan by new Democratic governance which both understands and is committed to the environment. An essential component to an effective solid waste plan in New Jersey is reduction of the waste stream by increasing recycling.[10]

The gubernatorial candidacy of James Florio included extensive consideration of "the waste issue," including a six-point plan for "getting out from under our garbage problem." In contrast to the proposal to site twenty-two new WTE facilities in the state, "The Florio Plan" focused first and foremost of waste reduction, setting a goal of recycling 50 percent of the state's waste stream, and enhanced public education about wastes enacted through New Jersey's public schools. Florio's plan bluntly rejected the incineration proposal, explored later in this chapter, asserting that "we don't need 22 incinerators . . . *While [we] review our current strategies, no new permits will be issued for incinerators*."[11] Leveraging the strong environmental record he amassed over years as a member of the U.S. House of Representatives, Florio pitched to voters a "rational solid waste policy . . . based on sound economics, good science, tough enforcement and common sense. This plan is not only environmentally sound, but will save taxpayers millions of dollars as well."[12] Florio argued, "New Jersey needs a comprehensive statewide solution that emphasizes waste reduction and recycling. Our State faces solid waste problems that the rest of our Nation will soon face, and we have the opportunity to show the country how imaginative and environmentally sound policies can save money, precious resources and lives. We can solve the problem and under the Florio Administration, we will solve it."[13]

On the cusp on the 1990s, waste management in New Jersey had become something of a partisan wedge in the state's politics. But not long before, waste management had been largely a nonpartisan issue attracting common concern and attempts at collaboration. Why did the shift away from landfilling and toward either recycling or WTE—or some combination of both—now appear to rile partisan perspectives? This chapter examines the histories of both recycling and WTE in New Jersey and considers how and why each came to attract such controversy within and beyond attempts to resolve the state's waste management crisis of the 1980s.

Recycling in New Jersey

Though it will never be settled who exactly *invented* recycling, it is clear that a desire to repurpose the waste stream for some sort of beneficial purpose permeated into the public consciousness along with the broader awakening of environmentalism in the United States during the 1960s and 1970s. By the 1980s, recycling was attracting considerable attention from the public and policymakers alike. According to a number of reports, in 1980 the town of Woodbury (Gloucester County) became the first municipality in the United States with regular curbside collection of recyclables. Though, not all were pleased with the program as several residents protested the new protocol by dumping trash onto the lawn of the town's mayor, Don Sanderson. Securing the compliance of town residents took some time, but within a few months over 80 percent of town residents were reportedly participating.[14]

Elsewhere in New Jersey, recycling began in an ad hoc fashion, oftentimes on the backs of volunteers and civic organizations. Wayne DeFeo, a former solid waste official with both Somerset County and the New Jersey Board of Public Utilities, recalled in an interview the early years of recycling in Somerset County:

> We were doing curbside [collection] . . . we used, at that time, you would look at a bread delivery van . . . Take that van, big open back with postal hampers and three people and you were stacking newspaper and throwing bags in postal hampers. No radios, we didn't have anything. We were just running a route, working with people who develop routes . . .
>
> We were sorting by hand. We were using small equipment for recycling . . . And we did not have equipment, we did not have trucks, we did not have anything . . . We started to develop . . . a conveyor system. Nobody had these systems here, so we developed with our engineering firm that we had had on retainer, a mobile conveyor system. A conveyor system built in components that could bolt together in the morning and the feed conveyors coming in were a long line, went out overhead doors, because we were in an airplane hangar, effectively. Went overhead doors into bins. At night we would drag the whole thing back in, because we couldn't store it outside . . . We went from sorting on card tables to having . . . literally, what was it, 30-foot ceilings, they were floor to ceiling bags of recyclable. Floor to ceiling with paper. We couldn't move it out, because we didn't know how to process it.[15]

In the early 1980s, as recycling ramped up due to citizen and municipal interest, the State Department of Energy and the NJDEP formed a joint New Jersey Office of Recycling. This office, intending to consolidate state action around recycling, worked with the newly formed New Jersey Advisory Committee on Recycling, an entity tasked with shaping the state's foray into this new dimension of solid waste management. The Committee on Recycling, like the broader Solid Waste Advisory Committee formulated through the 326 amendments, had a membership comprising environmentalists, government officials, and representatives from businesses and industry sectors perceived to be stakeholders in recycling and materials processing operations. In 1980, the Committee on Recycling along with the NJDEP and the Department of Energy released New Jersey's first state plan for materials recovery, called *Recycling in the 1980s*.[16]

The Committee on Recycling encompassed a broad range of viewpoints and philosophies, yet coalesced around several key recommendations to state legislators. These included a "landfill disposal surcharge," which would establish a funding source for a recycling trust fund aimed at supporting emerging municipal recycling programs like those in Woodbury and Somerset Counties; a 50 percent recycled products procurement directive for state government and agencies along with development of procurement best practices for businesses; creation of public relations, advertising, and K-12 educational programs to build public support for recycling; adoption of tax incentives for recycling processing firms and the beverage and container industry; and restrictions on "flow control" directives aiming to prevent disposal of recyclable materials in the state's landfills and incinerators.[17]

Recycling in the 1980s proposed a five-year program for achieving a 25 percent or better recycling goal; that is to say, by 1986, 25 percent of all the waste generated in New Jersey each year would be diverted for materials recovery. Put another way, the goal was to essentially reduce the amount of material flowing to state landfills by 25 percent. The recycling plan targeted an ambitious range of materials and substances, including paper, metals, glass, plastics, oil, tires, food, and yard wastes, equating to about 1.3 million tons of material.[18] According to Mary T. Sheil, one of the first directors for the Office of Recycling, the purpose of setting such a goal for 1986 was to:

1   Decrease materials flow to overburdened landfills.
2   Develop organized source separation programs that offer the potential to provide a stable source of raw materials supply to secondary material industries.
3   Reduce the design capacity of energy recovery systems [WTE incinerators] by 3000 to 4000 tons/day, for an estimated capital costs savings of $260,000,000.
4   Reduce solid waste management collection and disposal costs which are rapidly increasing because of the closing of landfills near population centers and requirements for stringent environmental landfill controls.
5   Conserve energy in the manufacturing process by at least 2 million barrels of oil equivalent per year.[19]

A year later, the state legislature passed as an amendment to the 1975 Solid Waste Management Act, the so-called 1981 Recycling Act, which aimed to establish one of the primary mechanisms for achieving the five-year recycling rate target: the "landfill disposal surcharge." The preamble to the 1981 law notes a clear connection between "energy, environmental and economic problems," and points out that any solution to this triad of challenges "requires proper solid waste and resource recovery management."[20] More specifically, "The Legislature further finds that the recycling of waste materials decreases waste flow to landfill sites, recovers valuable resources, conserves energy in the manufacturing process, and offers a supply of domestic raw materials for the State's industries; that a comprehensive recycling plan and program is necessary to achieve the maximum practicable recovery of reusable materials . . . and that such a plan will reduce the amount of waste to landfills."[21] The 1981 Recycling Act articulated the landfill disposal surcharge and the associated trust fund, which were intended to kick-start county and municipal recycling efforts. The act imposed a twelve cent per cubic yard charge on all waste disposed in the state's landfills, to be assessed and collected by the landfill operators. The funds collected from this tax on landfill disposal would go into the new State Recycling Fund, established as a nonlapsing revolving fund jointly administered by the NJDEP and Department of Energy.[22] Initial estimates suggested the fund would hold between $4 million and $5 million annually,[23] with monies being directed to municipalities, recycling businesses, and state agencies for the following purposes (per the original legislation):

1 Recycling grants to municipalities. Local governments are reimbursed on the bases of the quantity of materials recycled . . . or the basis of increase in that volume. Not more than forty-five percent of the tax receipts are to be used for the grants.
2 Low interest loans and reserve fund for a loan guarantee program for recycling businesses. Not more than twenty percent of the tax receipts are to be used for this program.
3 Public information activities. These are conducted by the Office of Recycling and are limited to fifteen percent of the tax receipts.
4 Program administration. County and municipal programs are allotted ten percent of the receipts, as is the Office of Recycling.[24]

The provisions of the 1981 Recycling Act were seen as a vital boost to a growing area of both civic and commercial interest. Sheil reports that before the adoption of *Recycling in the 1980s* and passage of the Recycling Act, there were about 250 recycling programs in the state (many unaffiliated with any particular unit of government); most of these "received little municipal support and were generally volunteer citizen organized activities rising out of the environmental movement of the early 1970s." Due to the limited nature of these primarily volunteer and civic organizations "the percentage of material recovered compared with the population served in most towns was minimal, often reflecting a very low participation rate."[25] However, immediately after the implementation date of the Recycling Act in January 1982, the number of recycling programs had increased to over 400 groups active in more than half of New Jersey's municipalities. Sheil reported that by early 1982, at least 165 of New Jersey's municipalities were offering their own curbside recycling collection services.[26]

Accordingly, the mid-1980s saw a flurry of activity surrounding recycling. During the 1970s, the State Department of Energy, NJDEP, and state Solid Waste Advisory Council had occasionally published guides for recycling aimed at municipalities, civic groups, and interested individuals, but this genre exploded in the 1980s.[27] After release of *Recycling in the 1980s* and passage of the 1981 Recycling Act, the Office of Recycling would come to publish dozens of reports, guides, and educational materials through the early 1990s. These ranged from instructions and guides to completing applications for the newly created State Recycling Fund grants, to directories of recycling outfits in the state, to best practices for handling, sorting, and processing different types of materials from motor oil to autumn leaves.[28] "Recycling Roundup" and "Recycle-Gram" newsletters offered readers updates on new programs and initiatives while informing readers about the contours of new and emerging programs and rules.[29]

Central to this literature, regardless of the specific topic of any particular publication, was the theme of market development. From the outset, recycling advocates in New Jersey realized that for any recycling program to be effective, a robust market of materials buyers had to be identified and encouraged. These were, after all, the *customers* that recycling programs were aiming to supply with material. Among the first state publications focused on developing recycling programs—a 1972 guide "to aid a community in setting up a recycling program"—informed readers that identifying and connecting with "secondary materials merchants" was "the single most important aspect of operating a recycling program and it is important to maintain a good business relationship with the purchasing dealer."[30] During the 1980s, the Office of Recycling published several guides to this topic, ranging from issues like generating publicity for recycling to ideas for writing contracts.[31] One study, released in 1988, offered a detailed quantitative assessment of existing and potential markets for a full suite of theoretically recyclable materials, including familiar products like paper and the rapidly growing category of plastic but also some unconventional products like tires and automobile components. This report noted that in the span of decade, New Jersey had become "clearly one of the leading states in solid waste recycling activities, providing a recycling industry not only for its secondary materials generated in-state, but also for secondary materials generated outside of New Jersey . . . Generally, the current initiatives [to encourage recycling programs] were positively rated and several additional strategies have been suggested to meet particular market development needs."[32]

While the Office of Recycling was busy supporting on-the-ground recycling programs, the state legislature also considered additional steps in shaping the market for recyclable materials. The most significant of these in the early 1980s was exploring a container deposit program, or "bottle bill." A state bottle bill had been contemplated since the mid-1970s, but received much greater scrutiny and attention in the context of a broader push for state support of recycling programs through the 1980s.[33] The basic logic of any bottle bill is to apply a deposit to designated containers—typically plastic, glass, or aluminum bottles and cans—that is included in the purchase price of the container. After using the product, consumers can return the container to a designated location and receive a cash refund. The bounty placed on each individual container, according to the logic of bottle deposit programs, incentivizes consumers into recycling these items as opposed to carelessly littering or tossing them in with the rest of the waste stream. At the same time, however, most bottle deposit programs impose obligations on retailers to accept returns and manage deposit refunds, as well as on beverage distributers and drink-makers (or their subsidiary bottling units) to develop so-called reverse logistics networks to transport and process the recovered materials.[34]

All of these issues came to the fore during a series of contentious hearings held by the New Jersey legislature. The bottle bill was positioned as a measure that would not only support recycling efforts, but also inspire New Jerseyans to litter less. Indeed, in his introduction to the 1983 New Jersey General Assembly's Forum on the Bottle Bill, Assembly Speaker Alan J. Karcher noted that "The bottle bill, as you know, is one of the most important matters now under consideration in New Jersey . . . There is nothing that gives us a sorrier image than the condition of our streets and public thoroughfares. I am reminded every time I talk about this subject that, image-wise, New Jersey must do something."[35] In consideration of the bottle bill, scores of experts testified on both sides of the matter. Many of those favoring the legislation argued for the impact a bottle deposit could have on reducing litter, but for the direct and immediate impacts of diverting materials from the state's landfills. For instance, Eleanor Gruber, testifying on behalf of the League of Women Voters at that same 1983 forum, noted that "Most people sitting here today would agree that litter and solid waste disposal are real problems in our State. The first is litter on our streets, highways, parks, beaches, and your front lawn. The second is more serious, because we have a real crisis in New Jersey. Landfills are bulging at the seams; many have already closed. Trucks by the hundreds per hour dump potentially reusable products. We area a wasteful society, encouraged to use throwaway products by companies vying for market share."[36] Hundreds of resolutions of support accompanied Gruber's testimony, including nearly 250 issued from various municipal councils, ten county Boards of Freeholders, over seventy mayoral petitions, and nearly twenty environmental, waste management, and agricultural associations.[37]

Others in favor of the bottle bill, including mayor of the Camden County town of Haddonfield, John Tarditi Jr., argued that diverting bottles and cans from the emerging curbside recycling collection schemes would create space for still other materials, like yard waste, to be collected. "The problem in the past," he noted, "has been that rarely is a community in a position to collect these materials, because of all the time and money being spent on recycling beverage containers."[38] The secretary of the New Jersey chapter of the Sierra Club, Albert Kent, claimed that based on the group's own independent study, just 5 to 10 percent of all glass containers in New Jersey were being recycled. The bottle bill, Kent claimed, "will increase glass recycling by at least a factor of four, and still leave a couple of hundred thousand tons per year of non-beverage glass" to be managed.[39] Tarditi, Kent, and other proponents asserted that the bottle bill could be a boon for the struggling New Jersey glass industry and emerging recyclables processing industry, by spurring investment in new recycling and processing equipment while sustaining the demand for labor in this sector of the economy. Kent, for example, claimed that "the deposit law will create three to four thousand new part-time and full-time jobs in the receiving, handling and recycling of used containers. These will be mostly for unskilled, young, presently unemployed workers. The pay will be low, but the jobs are desperately needed."[40] Still others appeared before the legislative committees to argue that the bottle bill would reduce the amount of broken glass in public parks (protecting playing children and works employees alike) and reduce the unsightliness of the streets around the Meadowlands sports complex (along with Yankee Stadium, despite being in another state).

Wide-ranging support for the diverse arguments in favor of the bottle bill did not, however, mean that the measure—or even the concept—was accepted by all. Indeed, the bottle bill encountered staunch opposition from representatives of key economic sectors. These included food retail groups, glass and packaging manufacturers and their labor unions, and drink-makers and bottlers. One of the loudest sources of opposition to the bottle bill was the New Jersey Food Council, a trade association representing the retail grocery business in New Jersey. The position of the food council was clear—the bottle bill would kill New Jersey's grocery industry. Barbara McConnell, the president of the food council, testified: "Ask any food retailer in any state that has a bottle bill, and he or she will tell you that is it an absolute nightmare and that it is one of the most costly, inefficient, and unsanitary systems one can imagine."[41] McConnell cited analysis by her group and similar entities around the country, that bottle bills would require "800–1000 additional square feet of space" for collection and sorting purposes, requiring stores not only to reduce their sales space but also invest in new infrastructure (registers, storage bins, pallets, etc.) that did not contribute to sales. But this was not the end of the analysis. McConnell noted that the bill required retailers to refund deposits for any product they carry, even if the customer did not purchase a particular container in a particular store, "which means in some states that supermarkets are redeeming 200% to 300% over what they sell . . . causing them to put money up front that they never collected."[42] She cited a time, "not long ago that public health and government officials were calling for an end to the returnable container system because of the potential health hazards they generated."[43] And of course, all of this would require additional labor and employee hours to implement.

Others at the hearings would argue that each of these issues could be offset through careful policy design, for instance subsidizing equipment necessary for retailers to implement the bottle bill, extending to retailers a "handling fee" of some sort, or specifying the conditions under which retailers would be obligated to provide a refund. But McConnell's testimony cut to some of the more fundamental claims associated with bottle bills as well:

> To impose that kind of cost on the consumer of this State through a system that will deal with less than 5% of the solid waste and less than 20% of the litter [the estimated amounts of the waste stream comprised of containers affected by the bottle bill] is irresponsible.
>
> First, the proponents of a bottle bill claimed it would save enormous energy and natural resources. Then they supported an amendment of the bill to make it a returnable, rather than a refillable system, which negated the energy saving argument and also was an admission that job losses could possibly occur under a deposit system.
>
> Then they argued it would solve our shrinking landfill problem, although bottles and cans represent only 5% of the garbage that is landfilled in New Jersey. A $100 million solution to a $3 million problem hardly seems logical to me. Now the proponents claim a bottle bill will solve our State's litter problem and will change the sloppy habits of people who litter . . . You can't deal effectively with any problem by attacking only a part of it, and our State's problem is only partially a matter of bottles and cans.[44]

McConnell's was not the only testimony mobilizing these types of arguments. Representatives from the United States Brewers' Association, the Association of New Jersey Convenience Stores, Pepsi Cola Bottling Company, and various liquor and beverage distribution companies all made similar cases. Many, including McConnell, argued that instead of implementing a law focused on such a narrow slice of the waste stream, the state should direct more support for municipal recycling programs through the provisions of the 1981 Recycling Act.

Even some key stakeholders in the state's oversight of recycling were unenthusiastic about a bottle deposit scheme. For instance Commissioner Leonard S. Coleman Jr. of the State Department of Energy—one of two agencies behind the Office of Recycling—argued that "deposit legislation should not be designed to remove an important element of the waste stream from local recycling programs, but rather should assist and simplify these growing programs by providing incentives so that more people participate in them."[45] Commissioner Coleman made the case that individuals motivated to recover the deposit on glass and metal beverage containers would divert revenue away from the towns and civic groups powering the growth of recycling in the state. Sensing that danger of anything which could potentially undercut the growth and stability of municipal recycling programs in the state, Commissioner Coleman reminded those gathered at the forum "that a crisis in solid waste has been growing in New Jersey for the last ten years . . . We must do all we can to reduce the volume of solid waste disposed of in New Jersey by promoting the further growth and success of our recycling program. Any deposit legislation must not represent a 'quick fix' approach to a complex and easily misunderstood problem, but must be part of an integrated and comprehensive effort to reduce the waste stream in New Jersey."[46]

The commissioner's remarks also moved the conversation beyond the world of waste management, by connecting the bottle bill to industrial and labor concerns that had been brewing in New Jersey for several years. In addition to the impacts on recycling and waste management more broadly, Commissioner Coleman asserted that "perhaps more importantly, deposit legislation in New Jersey must not have an inimical impact on the State's glass industry." He continued, "We cannot afford to allow any new program have a negative effect on employment in an industry which has already suffered greatly as a result of changes and recession. In fact, if there is to be deposit legislation, it should be carefully tailored to benefit the industry by providing economic advantages and cost savings. Any deposit program should make glass cullet [processed recycled glass] available to glass companies in New Jersey at extremely low cost, thereby providing a significant economic boost to the industry . . . This part of the program would complete the recycling chain and enhance the position of our State's glass companies."[47]

Commissioner Coleman was not alone in explicitly examining the labor and industrial impacts of the bottle bill. Representatives from New Jersey's AFL-CIO labor union as well as some of the state's glassmaking firms testified at the 1983 forum and hearings in subsequent years. Robert C. Donovan, who was the associate public affairs manager at the Owens-Illinois Glass Company with manufacturing facilities in southern New Jersey, testified that "From a business point of view, we oppose deposit laws because of their detrimental impact on the beverage container industry and the persons employed within it." He noted, that "New Jersey, as you know, is the home of a large glass container industry, a can manufacturing industry, and a plastic products industry. Within those industries, we have sixteen beverage container manufacturing plants, employing more than 6,500 persons presently, with an annual payroll in excess of $150 million. The plants are responsible for the purchase of more than $200 million in raw materials, goods, and services from hundreds of small and large suppliers located in New Jersey and other states . . . In addition, more than 1,100 Teamsters are employed in the transport of raw materials and finished products by those companies and would be affected, again, in proportion."[48] Donovan claimed that in states that had experimented with bottle bills, actual consumption of the beverages affected declined considerably—in Massachusetts, for example, Donovan claimed that the bottle bill there caused beer and soft drink consumption to fall between 10 and 15 percent, respectively. "That is just a sampling," he said, "but it shouldn't take a mathematical genius to realize that a ten to twenty percent reduction in the beverage market translates into a ten to twenty percent reduction in container requirements, production, and jobs."[49]

Taking a similar tack was Charles Marciante, president of the New Jersey State AFL-CIO, who testified alongside the international vice president of the Glass, Plastic, Pottery, and Allied Workers Union, Joseph Mitchell. Marciante made an emotional appeal directly to the elected officials of districts with considerable glass and packaging industries. "We have to look at some facts . . . Going back to 1978, we had nearly 14,000 glass workers in our organization. In 1983, December, that membership has been reduced to 3,926 . . . This piece of legislation will hurt a lot of fine people."[50] Marciante continued, "You'll hear the argument . . . that this legislation requires that all bottles and so forth be returned, will be ground up, and the workers will not be jeopardized because they will have to make additional bottles to compensate for the bottles which have been ground . . . [but] the cost of the product goes up dramatically, to the point where a person instead of buying a six-pack, buys a single container. When you reduce that six bottles to one, you have a commensurate loss of production, and with the commensurate loss of production you have a commensurate loss of jobs. Those are the hard facts."[51] These assertions did not ring true to all of the members of the assembly present. Alongside consideration of the bottle bill across the United States was the rapid increase in the use of plastic bottles and aluminum cans as alternative forms of packaging products that had previously been delivered exclusively in glass. The following heated exchange between Assemblyman Arthur R. Albohn and Marciante illustrates the selective nature of the debate about glass containers and jobs:

ASSEMBLYMAN ALBOHN: That is not true, Mr. Marciante . . . you discussed in your remarks the number of glass workers over the years in the State of New Jersey, and I think you meant all the glass plants in New Jersey, not just one.

MARCIANTE: You are correct, sir.

ASSEMBLYMAN ALBOHN: So, the point is, that while you may be casting aspersions at the bottle bill legislation for the reductions at Owens-Illinois, the fact of the matter is, there are glass plants which do not have the major part of their production in beverage containers which are also closing down, or shutting down, and their announced reasons are . . . the high cost of utilities in the State of New Jersey, as well as the high value of the labor and benefits contracts.

MARCIANTE: That is indeed part of the problem, but, also, this particular piece of legislation will merely compound those problems dramatically.

ASSEMBLYMAN ALBOHN: All right. Then, let's just suppose you were asked the question, "What do you fee then is the greatest threat to the glass industry in the State of New Jersey?"

MARCIANTE: Assembly Bill 1753 [the bottle bill] right now. (Applause from audience)

ASSEMBLYMAN ALBOHN: More so than canned beverages?

MARCIANTE: Absolutely . . .

ASSEMBLYMAN ALBOHN: More so than plastic containers?

MARCIANTE: Yes.

ASSEMBLYMAN ALBOHN: Is this why Owens-Illinois has not cut back on its plastic container production, but has cut back on its glass production?[52]

Marciante and other industry representatives were hesitant to acknowledge anything other than the bottle bill as contributing to the loss of jobs and closure of facilities in the glass industry. Testifying again two years later when the assembly once more contemplated beverage container deposit legislation, he assured the policymakers present that "I am cognizant of the fact that a number of the major companies are shifting to plastic containers . . . [but] with the adoption of deposit legislation . . . as each state implements a deposit bill, there is a falloff in the number of jobs in the state of New Jersey . . . I am not talking about vast numbers of people, but even if it is one person, it is one hell of a big loss."[53] Acknowledging that growth in the collection and processing of recyclable materials was indeed producing new jobs in the state, Marciante mused to those assembled: "I wonder how they will feel when these jobs are lost and they can only go to all of the thousands of [recycling] jobs that will pay only $4.00 an hour? You try to get by on $4.00 an hour, or try to raise a family on $4.00 an hour for a 40-hour week. Why, it is damn near a joke. Yet, they are the kinds of jobs—and $4.00 might even be a little high—that are being put in to replace jobs which pay $12.00 and $14.00 an hour, jobs that people have negotiated for, with those kinds of wages, over the years."[54]

Immediately following Marciante's 1985 testimony, however, was James Lanard, the chief lobbyist for the New Jersey Environmental Lobby, a coalition of environmental advocacy groups in the state. Lanard read into the testimony an editorial from what was then the largest newspaper targeted to African American readers, *New Amsterdam*. Based in New York City, *New Amsterdam*'s editors reported that "The dozen recycling plants . . . are, in most cases, located in minority areas, providing jobs and creating business for ancillary services such as food shops, restaurants, gas stations, and other commercial enterprises . . . Over 2,500 jobs for minorities have been created already with the New York Bottle Bill."[55] Lanard, who was testifying in favor of the bottle bill, delivered a blistering critique of the claim that deposit legislation was ruining the glass industry.

> If I may quote from *Business Week*: A major glass company executive said . . . "We lost many millions of dollars of glass container volume to our own plastics division." That is a quote . . . and what that clearly reflects is that the glass industry has chosen to invest, quite heavily, in plastics . . .
>
> Now, let me give you some examples of three companies that had been located here in New Jersey and what their corporate practices are: Anchor Hocking Company divested its glass container division, but acquired Gibson Associates, a plastics manufacturing plant. Kerr Glass is beefing up its plastics packaging division, and it sold its glass plant due to plastics competition. Brockway closed a glass plant, but bought several plastics companies. Is this a commitment to New Jersey workers? Are these jobs being located here in New Jersey?
>
> Are these companies shutting down glass companies in the labor markets of New Jersey and opening up plastic companies elsewhere? That is clearly the case. I suggest to the Chairman and to the members of this Committee that if the glass industry is really concerned about jobs in New Jersey, those plastics plants of Anchor Hocking, Kerr Glass, Brockway, and Owens-Illinois would not be located out of state, but would use the same work force, trained and expert in making glass bottles, to shift their work to plastics. That has not occurred, and that is the real answer.
>
> It explains why the workers in New Jersey have been knifed in the back by the same people who are now professing to save their jobs.[56]

Assemblyman Robert P. Hollenbeck, who chaired many of the hearings and forums relating to bottle deposit law in New Jersey during the early 1980s, made clear that the employment argument against the bottle bill was specious at best. "I think the public sometimes gets misled by some people. The problem with the glass industry in the State of New Jersey stems from the competition by plastics. Guess what? It is your own company that is manufacturing plastics in other states. That is where the real problem comes from . . . What you are trying to do is defeat a piece of legislation, but you are not trying to solve your problem."[57]

But Assemblyman Hollenbeck went further, identifying plastics as the most fundamental determining factor in the success or failure of New Jersey's recycling programs in their entirety. In a series of very prescient observations, he explained: "Ladies and gentlemen, the problem you have, of course, is . . . plastic. We hear about recycling. It is a nice catch word . . . What about plastic containers, the enemy of the glass industry? The State, with all its solid waste problems, is going to be dealing with plastics, and it has no market for those plastics. There is no market for them. We don't know what to do with them. So, what are we going to do with all this recycled matter? We are going to wind up with landfill sites, under recycling, that are going to be filled with glass. We are going to have glass landfill sites. We are going to have metal landfill sites. And we are going to have plastic landfill sites."[58]

Hollenbeck's concerns centered on the missing linkage between the New Jersey's emerging recycling efforts and actual markets for the collected materials; he made the point that the state should only be concerned with materials that can be reliably sold and processed. In the absence of such markets, Hollenbeck argued for reusable materials to become the norm. In dealing with the perspectives on the bottle bill legislation, he was surprised that no one from the glass industry or otherwise made the case for reusable glass containers and refillable bottles as a more elegant solution that addressed the market issue and the employment issue in a single stroke.

New Jersey never did pass a bottle bill, and the 1980s bottle bill hearings suggest one of the reasons why. Once certain sectors and interest groups perceived that they could be negatively impacted by policies toward something as (formerly) innocent as waste management, they took a stand. In particular, the labor, industrial policy, and business rights arguments in opposition of bottle deposit law were clearly one of the avenues by which waste management became a political wedge in the state. Interest groups had cast an idea aimed at boosting recovery of glass—and successfully implemented in many other states and other countries around the world—into one which unfairly targeted blue-collar workers and dying industrial plants in poorer corners of the state. According to opponents like the New Jersey Food Council, the bottle bill would gut the food retail infrastructure urban and suburban New Jerseyans relied on for their meals.

On the other side of the table, many proponents of the bottle bill would voice their disapproval of the legislature's inability to pass a measure which would have increased recovery of one particular material and at least slow, though not solve, the landfilling crisis. Furthermore, though by the mid-1980s they had evolved into government-supported programs, many early proponents of recycling in New Jersey traced their origin to volunteer civic and environmental groups. For these folks, recycling measures in the state had not yet gone far enough. For those making an environmental case for recycling (as distinct from those making a case for recycling as a landfill preservation tool), the failure to implement the bottle bill was symptomatic of lawmakers' insincerity toward addressing the environmental dimensions of the solid waste management crisis.

Despite the flurry of activity pursued by the Office of Recycling, the dust kicked up in debates around the bottle bill, and the generally high praise New Jersey's recycling efforts received from evaluators and consultants located elsewhere, by the late 1980s the state still was not meeting the 25 percent recycling rate goal established in the *Recycling in the 1980s* statewide plan.[59] However in discussion of the bottle bill, it was clear that some progress toward the recycling goal was being made. For instance, State Department of Energy commissioner Leonard S. Coleman Jr. noted at the 1983 New Jersey General Assembly's Forum on the Bottle Bill, that

> The State's current recycling program is comprehensive, and I might add, that without question, it has been acknowledged to be the most comprehensive plan in the nation. It recovers all recyclable materials—paper, used oil products, food, and yard wastes, in addition to metals and glass. In addition . . . the recycling program includes an intensive education program to change existing habits and practices . . .
>
> [The recycling program] covers 4.6 million people in our statewide population of 7.5 million . . . The participating municipalities have been well rewarded for their efforts. This year [1983] the State distributed over $2 million in grants to 241 municipalities, based on the number of tons of material recycled by each municipality . . .
>
> The primary goal of the [1981] Recycling Act is to encourage recycling through positive economic incentives. As we are seeing, this approach has shown significant results . . . During calendar year 1982, over 260,000 tons of material were recycled by the 400 recycling programs operating in . . . the State's municipalities. To put that figure in perspective, the 260,000 tons are enough right now to fill Giants Stadium, one and a half times over. The goal of the program, if we achieve the 25% in another couple of years, would be to fill the stadium six times over. I'm sure it is the hope of all of the that the Giants will have a winning team at that point, and we won't need to cover it over six times. Nonetheless, the program has, in that sense and in how it has gotten off the ground, been very successful.[60]

But clearly by the mid-1980s, considerable challenges remained. The 1985 draft update to the state solid waste management plan estimated that even if every county could implement, immediately, the facilities and infrastructures mandated by the 326 amendments to update their disposal capacity and also achieve the 25 percent recycling diversion rates called for in the 1981 Recycling

Act, all but four counties would still be facing shortfall in disposal capacity. That is to say, seventeen of New Jersey's twenty-one counties would potentially have nowhere to dispose of their trash, and most counties facing a shortfall were anticipating missing their target not by a few thousand tons, but rather by millions of tons of trash each by the year 2000.[61]

In attempt to avert this crisis, in 1985 the legislature began consideration of mandatory recycling laws. Owing to its grassroots origins, to this point recycling in New Jersey had been a voluntary affair, with towns deciding whether to offer recycling collection services. While some towns had already made recycling a part of residents' lives through regular curbside collection, others had ignored recycling almost entirely. This would change with the introduction of legislation to require all New Jerseyans to recycle, "a program more ambitious in its scope than anyone dreamed possible a few years ago" in the words of Assemblyman Albohn, the new law's initial sponsor. He would continue, in introducing the legislation,

> Recycling has now come of age. It is no longer a tool of hobbyists, lay environmental enthusiasts, or for the casual attention of homemakers, scout troops, etc. On the other hand, we owe such individuals our gratitude also for laying the groundwork for recycling on a far more extensive scale. They persisted in spite of a variety of obstacles, the major one of which was disinterest on the part of the public in general.
>
> But the problems faced in our State with the disposal of solid waste have now forced recycling into the center stage limelight . . . We can no longer tolerate disinterest, apathy, or annoyance at minor inconveniences that may have to be tolerated in reducing the magnitude of our solid waste disposal problem.
>
> As a result, [the bill] is a hard-hitting, tough bill. It pulls no punches. It calls for commitment and involvement on the part of everyone, every citizen, every industry, every business operation, every level of government. Such commitment is absolutely necessary if we are to resolve the problems the State and nation face, not simply with hazardous waste, industrial waste, toxic waste, and sewage waste but with the everyday products of our everyday lives—the waste from our homes and our style of living, whether it be humble or grand.[62]

The mandatory recycling laws were indeed tough bills, in the sense of the obligations they imposed on communities, individuals, and businesses. The original pieces of legislation proposed in both the assembly (A-3382) and the senate (S-2820) required counties to add specific recycling plans to their overall waste management plans that had been required under the 326 amendments years earlier and also appoint a county recycling coordinator. The laws required that the new county recycling plans designate "the recyclable materials to be source separated in each municipality, which shall include at a minimum aluminum beverage containers, and at least two other recyclable materials separated from the municipal solid waste stream" as well as a "strategy for the collection and disposition of source separated recyclable materials in each municipality" and recycling targets of at least 25 percent.[63] This meant, in effect, that except for aluminum, counties and towns could each be separating different materials from one another.

The logic behind this hyperlocal designation of recyclables was to allow flexibility in response to local markets for materials. The focus on forging connections between collected materials and demand for those materials was so important that the recycling laws not only required collection services, but also demanded counties to find buyers for the materials collected. In particular, the laws required that "Each county shall, within six months of the adoption of the district recycling plan . . . solicit proposals from, review the qualifications of, and enter into contracts on behalf of municipalities with persons providing recycling services or operating recycling centers for the collection, storage, processing, and disposition of recyclable materials designated in the district recycling plan in those instances where these services are not otherwise provided by the municipality . . . Each county shall continue to solicit recycling services as may be necessary to achieve the maximum feasible recovery targets."[64] While there were provisions for relief in the circumstance of no purchasers being available, from the outset the mandatory recycling laws focused counties' attention squarely on the materials that already had a market. The laws required municipalities to pass ordinances requiring residents and most businesses to "source separate" materials being collected at the curb, but also demanded action in several other important areas. For instance, the laws also required any firm that had previously been granted a license to collect wastes by the state to now offer recycling collection services. The laws required municipalities to collect fallen leaves and obliged landfills to refuse to accept fallen leaves. The laws required that containers sold in New Jersey have some sort of marking indicating their material composition, as to facilitate recycling, and obliged state purchasing agents to select products made from recyclable material whenever feasible. Finally, the laws confirmed the state's commitment to tax disposal at landfills and redistribute these monies to municipal recycling programs—as well as recycling processing companies—and also proposed new tax and financial incentives for companies to purchase products made from recycled materials.

Naturally, such a sweeping set of proposals attracted great commentary in the legislature. George Tyler, then an assistant commissioner at the NJDEP, described the mandatory recycling legislation as "vital and essential for the long-term environmental and economic well-being of New Jersey."[65] Tyler went on,

The natural question this bill provokes, is . . . why mandate recycling? As you know, since 1982, with the adoption of the Recycling Act, we have been operating a voluntary statewide recycling program. We have found this program to be exemplary; it has functioned well and has met with notable successes. Indeed New Jersey's program has gained national attention and recognition . . . We estimate that a million tons of waste is now being recycled in this state. This represents about 15% of the municipal waste stream and 10% of the total waste stream. But this is not enough . . .

To conserve existing landfill capacity and to avoid the costs associated with disposing of waste in expensive new landfills and waste to energy facilities, we must restructure management of solid waste in New Jersey. Increased recycling must be the cornerstone of our waste management strategy. The excellent start we have made must be built upon and expanded. It is imperative from an environmental perspective. Happily, mandatory recycling can be highly beneficial from an economic, energy management, and employment viewpoint as well . . .

Recently, when the courts were confronted with the problem of dwindling landfill space in specific areas of the State, they began to order entire counties to institute mandatory recycling as a means of conserving what little space is left. This judicial logic parallels our thinking . . . Our experience in New Jersey tells us the highest recycling rates are achieved by those communities that: 1) mandate source separation; 2) provide convenient collection systems; and 3) perform effective education programs. Based on that experience, we believe recycling should be mandatory statewide in order to achieve our waste management goals.[66]

Many others made similar cases in support of the mandatory recycling legislation in the state assembly as well as in support of the companion legislation before the state senate.[67] As with the bottle bill, however, not all who appeared before lawmakers were supportive of the new measures. For instance, Linda Pelrine, representing the New Jersey State Chamber of Commerce, was concerned about the impacts of mandatory recycling on existing recycling services. Because the laws directed counties and municipalities to develop recycling services so quickly, she anticipated that this work would be farmed out to towns' public works departments as opposed to private contractors. Pelrine and others feared that "these proposals do not protect the private sector in its efforts to continue established recycling programs or begin new ones," and she asserted that "the fundamental right to compete must not be denied to commercial, non-profit or for-profit organizations by new municipal programs . . . the main purpose of the legislation should be to encourage new and additional collection and reuse," not displace the efforts of existing recycling service providers.[68] More fundamentally, however, Pelrine and the New Jersey State Chamber of Commerce took "strong exception to the absolute power given to the Department of Environmental Protection ... The enforcement of these particular provisions of the recycling bill would amount to virtually the most offensive, costly, and ineffective consumer tax ever imposed upon the citizens of this State. It is, in every sense, an overkill."[69]

A common refrain among those testifying in opposition to the mandatory recycling provisions was that the laws tried to impose too much, too quickly, trying to force the collection of materials like plastics for which no viable markets existed. Furthermore, many of those opposed to the new rules felt that mandatory recycling would immediately come into conflict with the emerging plans for a statewide network of WTE facilities, discussed in the next section of this chapter. David Nalven, testifying on behalf of the New Jersey Business and Industry Association (one of the largest associations of employers in the state, then and now), submitted the following comments to the assemblymen present:

> Recycling in New Jersey may well be an idea whose time has come. We do have a crisis in available landfill space, and we have been wasting valuable resource by burying them in the ground ... However, implementation of a workable statewide recycling system will not happen overnight, and it won't happen by government decree. The legislation being considered today, in its efforts to be comprehensive—to be a single answer to a series of different and not necessarily similar policy questions involving not only the disposal, but the manufacture, sale, and use of paper products, bottles, cans, plastics, rubber products, and so on—ends up creating more problems than it would solve.
>
> In our view, the legislation fails to recognize that there are limits to the amounts of recycled materials for which there are short-term markets, and that those markets will expand only gradually—not by government fiat. There are active private and public recycling efforts already underway that this bill would trample ... I believe most members of our Association support the concept of a statewide recycling program and will do their share in trying to make whatever system is developed work. However, I believe that we are being very naïve if we think we can build a workable recycling system without adequate markets for the material we recycle. You cannot recycle a material if you can't sell it or give it away ...
>
> This bill is an attempt to define and implement a comprehensive recycling system for the State. Unfortunately, the interrelationship of the various regulatory, market, and political components hasn't been well thought out. The bill places requirements on existing county solid waste management systems that may be counterproductive. This is not a cohesive program.[70]

Over these objections, and after an additional year of hearings, in 1987 the legislature passed the very first mandatory recycling laws in the United States.[71] Even as great uncertainty swirled around exactly how the new rules would work in New Jersey, many including the state legislature noted the symbolic importance of a law requiring such close attention to the waste stream. For instance, in the preamble to the statute, lawmakers wrote, "The Legislature therefore declares that it is in the public interest to mandate the source separation of marketable waste materials on a Statewide basis so that reusable materials may be returned to the economic mainstream in the form of raw materials or products rather than be disposed of at the State's overburdened landfills, and further declares that the recycling of marketable materials by every municipality in this State, and the development of public and private sector recycling activities on an orderly and incremental basis, will further demonstrate the State's long-term commitment to an effective and coherent solid waste management strategy."[72]

There was considerable uncertainty regarding how the new rules would function in practice, and to whom (and at what price) recovered materials would be sold. In particular, the 1987 market development study produced by Office of Recycling contractor Arthur D. Little noted a range of barriers facing materials recovery in the state:

> Technical—including raw material specifications for products in which recyclable materials can be used. Other technical barriers included lack of viable technology to separate and process these recyclable materials into marketable products, as well as difficulties in retrofitting currently used technologies to accept secondary materials [another name for recycled materials].

> Economic—including the net cost of using secondary materials. These costs include collection, transportation, and processing costs. Costs of equivalent primary or other secondary materials were also reviewed to determine the competitiveness of secondary materials.

> Social/Institutional—including barriers to collecting and using recyclable materials. These barriers include consumer reluctance to participate fully in certain recycling programs, consumer bias against products manufactured from secondary materials, institutional encouragement of primary material usage, and restrictions on the use of recyclable materials based on government or trade association requirements and standards.[73]

Even studies from Office of Recycling expressed doubts about the ability of New Jersey markets to absorb the flow of recovered materials. Administrator Sheil observed that "stronger markets in some areas are necessary to achieve the goals of the state plan." While aluminum and paper markets were fairly stable, "glass, a previously stable commodity, is losing market demand to plastics. Paper, ferrous metals, plastics, and rubber require expansion or development to achieve the state recycling goal . . . a greater commitment by industry to the utilization of New Jersey generated material is required."[74]

Many commentators had noted in the lead-up to the mandatory recycling requirements, that some parties would serve to benefit much more than others from the new rules. David Nalven of the New Jersey Business and Industry Association envisioned a scenario where "transporters and warehouses of recycled material will be getting rich without producing any gain to the local community." He continued, "I can envision a scenario that has a homeowner sorting his trash, which is picked up by collectors, taken to a collection point where it is resorted and checked, and then taken to an interim storage facility pending sale to a recycler. Finally, without ultimate buyers, once again, it could be transported to a landfill or a resource recovery facility for disposal."[75] Others noted that a major barrier to enacting the mandatory recycling rules was the limited availability—and additional expense—of providing a second curbside collection to New Jersey's homes and businesses. Sheil, a director of the state Office of Recycling observed in a 1986 study that "the development of privately run collection routes requires new equipment purchases by industry as well as a recognition by municipalities that collection of recyclables will require contracts that not only benefit the community but also provide a reasonable profit for the collection industry."[76] She warned that "municipalities must also accept recycling as a solid waste management system that will vary in costs and benefits depending on local market conditions, availability of materials," and similar factors, and that "municipalities must view recycling not as a moneymaker but as a method to contain solid waste costs and to reduce waste to landfills."[77]

Some counties were already prepared to meet the law's obligations. Camden County, for example, passed its own mandatory recycling ordinance in 1985, which obliged all thirty-seven towns within Camden County to recycle newspaper, aluminum, and scrap metals along with yard and landscaping wastes. Mandatory glass and other metal container recycling followed in 1986, the same year the County debuted the Camden Recycling Facility. This was a joint venture with two glass and metals processing firms in the county, funded in part through the state grantmaking program implemented in the 1981 Recycling Act. By the late 1980s, Camden County had reduced the amount of material flowing to disposal facilities by approximately 40 percent. Municipalities' interest in complying with the new recycling ordinances and delivering material to the Camden Recycling Facility was also prompted, at least in part, by the nearly 300 percent increase in landfill fees at the facilities typically utilized in the county. As such, county and state officials alike were able to sell the success of the recycling program in terms of the tremendous cost savings it represented over sending materials to the landfill, and by the late 1980s estimated that each ton of recycling diverted in Camden County was worth about $65 in avoided tipping fees.[78]

However, many municipalities and counties were not pleased with these new obligations, especially as the questions of landfill and WTE facility siting remained, for the most part, unresolved. In its final form, the mandatory recycling laws obliged each county to formulate plans for recycling and have it approved by the NJDEP, appoint new recycling personnel, and enter into contracts with buyers of material. The laws mandated municipalities to similarly appoint dedicated recycling personnel and develop new collection systems while also obliging municipalities to pass new ordinances requiring "generators of municipal solid waste" to source separate designated materials, develop and distribute educational materials, and develop a plan for leaf collection—all within six to twelve months of the effective date of the mandatory recycling laws.

On the other hand, waste haulers saw a bonanza of additional routes and new fee structures. While the waste collection business had been divvied up for years (whether legally through the Board of Public Utilities or illegally through the actions of organized crime), recycling represented a new frontier for haulers and an opportunity to essentially double the amount of business they were doing in the state. Such was the anticipation for mandatory recycling that Sheil reported that even prior to the passage of the new laws, the New Jersey Waste Management Association was "gearing up for a campaign to educate its members on the need to change waste management practices in the State"—in order to ensure that the public demanded new recycling services for their town.[79]

Some of the provisions of the mandatory recycling laws were intended to expedite the development of recycling infrastructure in the state. For instance, local public contracts laws were amended to allow for the sale of recyclable materials outside of the regular public bidding process. Proposals for new recycling centers would not be required to secure "registration statements, engineering design approval, or an environmental and health impact statement;" conversely, no licenses for new disposal facilities would be granted unless the applicant could demonstrate to the satisfaction of the NJDEP and the Board of Public Utilities that the goals of the county recycling plan were also being addressed.[80] The flexibility was seen as a necessary slashing of red tape in order to get the recycling industry up, running, and hopefully profitable.

It is perhaps not surprising, then, that the emergence of recycling as a core activity in New Jersey's solid waste management system also presented itself as a tremendous opportunity for new scams. For instance, using "recycling" as cover for ducking the waste flow requirements imposed earlier was a popular tactic. A 1989 study conducted by the State Commission of Investigation observed an emerging strategy for misreporting recycling activity: "In pursuing waste flow violations the BPU and DEP have encountered the claim that the trash is merely being 'recycled' rather than shipped in violation of waste flow directives. However, after some recyclable material has been removed, remaining or residue waste is often shipped directly out-of-state, with huge amounts bypassing the authorized transfer stations. Thus, sham recycling operations severely complicate the job of enforcement."[81] The A-901 provisions discussed in the previous chapter focused on rooting out organized crime elements from the waste collection and disposal sector, but did not specifically address the new and substantially less regulated recycling industry at all. As a 2011 study of organized crime in the waste sector reported, "When New Jersey adopted the Statewide Mandatory Source Separation and Recycling Act nearly 25 years ago, the consensus was that recycling was not a lucrative enterprise and that incentives should be offered to encourage legitimate solid waste businesses to invest in the diversion of recyclable materials from landfills and incinerators. Those incentives included tax breaks and the deregulation of collection tariffs charged to solid waste haulers. Furthermore, the statute did not mandate new or additional licensing or background-check requirements for those who limit their industry involvement to recycling."[82] However, as recycling became established as a core component of New Jersey's approach to waste management, "this hands-off regulatory construct has not kept pace with changing economic trends and the opportunities they provide for both legitimate and criminally tainted business interests."[83] As the Commission of Investigation reported in 2011, the recycling industry "remains open to manipulation and abuse by criminal elements that circumvent the State's existing regulatory and oversight system. The urgency of this matter is compounded by evidence that convicted felons, including organized crime members and associates, profit heavily from commercial recycling, which, though a lucrative adjunct to solid waste, has remained largely unregulated. That is the case even though recycling has developed and grown to be an economic force far beyond what was envisioned when New Jersey adopted mandatory recycling nearly 25 years ago."[84] Jumping forward, briefly, from the late 1980s to the more recent past, New Jersey would find itself at the center of an increasingly global illicit waste trade centered around various recycling practices. While waste could not be stored for any substantial amount of time, recycling could be stockpiled, sold through opaque arrangements, passed off as better (or lower) quality than it really was, or even shuffled over to a landfill or incinerator as prices, time, and convenience dictated. As the commission reported in 2011, "Commerce in recycled paper, for example, is now international in scope, and it continues to mature at that scale despite cyclical price and supply volatility. Similarly, a significant global 'e-waste' market is emerging for recycled computer components and other electronic debris, some of it highly toxic. Meanwhile, vanishing landfill space and the increasing reclamation and redevelopment of 'brownfields' have put a financial premium on the collection, disposal and re-use of contaminated soil and other debris."[85]

Returning to the 1980s, the crisis in landfill capacity spurred dramatic action in development of landfill alternatives, and in the minds of many policymakers recycling was an immediate and relatively uncontroversial strategy for prolonging landfills' working lives. It was in tune with the public's calls for environmental protection and at least held the prospect of offsetting some of its own costs through the sale of the materials collected. In introducing another, extensive hearing on the state of New Jersey's recycling infrastructure, Assemblyman Hollenbeck once again offered a clear-eyed assessment of the problems:

> Trying to reduce our solid waste stream is something that has to be done because we are running out of landfill space and areas to take care of our solid waste. These problems would occur in a State like New Jersey first because of our dense population . . . [also] we just can't keep opening up new landfill sites throughout the State, since there is a potential threat to the groundwater supply.
>
> With the advent of resource recovery units [WTE] being installed and contemplated around the State, most people realize that they are extremely expensive to install and to operate . . . If we can reduce the flow by 25% by taking recyclables out of that flow, we can reduce the size of those particular plants . . . So, the need to do it to save the municipalities the cost of going to a resource recovery unit is very important. We realize that an awful lot gets into our landfill sites and our solid waste stream that really shouldn't be in there, including leaves in the fall and back yard waste. These things should really not be going into our landfills. These types of things should all be addressed. That is the purpose of this hearing.[86]

For many public officials at all levels of government, removing materials from the waste stream to reduce the flow headed to WTE or landfills was the most immediate and feasible solution to the disposal capacity crisis. But as Assemblyman Hollenbeck noted, recycling would have to work in tandem with WTE in order to fully address New Jersey's waste management problems.

Waste-to-Energy in New Jersey ("Resource Recovery")

Burning trash as a disposal strategy, like burying it in the ground, is a time-honored human tradition. In addition to dumps on the outskirts of towns, earlier generations also burned wastes of all types when necessary. Throughout the United States and around the world families have burned their domestic wastes in backyards, sometimes in hearths or outside on simple piles, in metal boxes or drums. Sometimes the strategies of dumping and burning were combined. Until the 1950s and 1960s, for instance, many if not the majority of town dumps in the United States would conduct regular burns to try and control odors and vermin. As the United States urbanized starting in the nineteenth and twentieth centuries, new multifamily dwelling units would utilize small, on-site furnaces meant for the occupants of a single building. These on-site furnaces and incinerators were convenient for residents and cities alike. At the same time, larger cities in the United States and Europe built neighborhood incinerators as well. Remnants of these structures and their smokestacks can still be spotted in many older urban cores.

The 1972 Musto Commission reported that there were about 6,400 of these residential incinerators and about forty neighborhood incinerators in New Jersey at the end of the 1960s.[87] Most of these were located in urban northern New Jersey. In New Jersey as elsewhere, these older types of urban incinerators were inefficient, dirty, and polluting. The furnaces found in many old-style incinerators did not burn at a high enough temperature to quickly and completely destroy the waste material, leading to volumes of ashy smoke that could coat neighboring buildings, streets, and parks. By the same token, incinerators remained attractive because they disposed of waste in a way clearly distinct from landfills and recycling. By burning wastes, the volume of material could be substantially reduced, lifting some of the pressure on dumps. The ability to dispose of waste in one's own residential building reduced the need for waste collection, and disposal at neighborhood incinerators minimized distances for hauling. Although incineration left much to be desired, it also represented an attractive and immediate disposal strategy for many growing cities.

In the 1950s, Swiss engineers improved on incineration in a way that kept the volume-reducing benefits while also improving the combustion performance and, therefore, environmental performance of the technology. By lining the interior of the furnace with pipes carrying water, and finding ways of burning wastes at a higher temperature (perhaps by partially drying the waste first, or augmenting the blaze with additional fuels), new water-wall style incinerators could also generate steam that could be used for industrial purposes, or linked to a turbine to produce electricity. Thus, energy-from-waste incineration was born. When the technology crossed the Atlantic, it generally became known as waste-to-energy (WTE) or euphemistically described as "resource recovery" and WTE plants as "resource recovery facilities."[88]

In New Jersey, as in many places, old-style incinerators started to fall out of fashion after World War II. Many built prior to the war had reached an advanced age and fallen into a state of disrepair. In other cases, communities living near incinerators complained about the smoke and debris. By the late 1960s, those incinerators that remained were coming under attack from environmental advocacy groups because of the air pollution that they generated and the associated threats to human and environmental health they represented.

As the composition of the waste stream changed to include more synthetic materials and plastics, the human health threat only grew as Americans came to realize that burning plastics at a low temperature was a recipe for releasing various toxins and carcinogens. In a 1965 hearing, the state commissioner of health Dr. Roscoe P. Kandle, testified to those present that "Eleven incinerators have been constructed in this metropolitan area [northern New Jersey] but only two are now in operation. Nine incinerators were abandoned or demolished because of the high cost of operation, repair and replacement. The two operating incinerators do not have the capacity to burn the refuse produced by these two municipalities, that is, the part of it which can be dealt with by incineration. They are Hackensack and Jersey City. As a result, the excess is brought to the existing refuse disposal areas located nearby."[89] Incinerators were seen as dirty, and too costly to repair in light of the possibility of sending wastes to dumps and landfills. More critically, these smoke-belching behemoths seemed backward in light of emerging recycling technologies and even in comparison to the new generation of carefully engineered sanitary landfills. The 1972 Musto Commission report observed that "Only *three* of New Jersey's 38 municipal incinerators are in operation today, and they do not meet new air pollution requirements of the Department of Environmental Protection. The major reason the 35 other incinerators have been closed is that they failed to meet design criteria and emission standards."[90]

Nevertheless, as Dr. Kandle's own testimony made clear, the technology offered something valuable in its ability to reduce volumes of waste. Indeed, as Dr. Kandle noted, "both incineration and landfill operations are always required; that is, these are not alternatives, they are supplemental methods."[91] There was hope that new incinerators could perform better and meet more stringent environmental requirements. Belief in the technology was so strong that, the first official state plan for waste management in New Jersey, released in 1970, featured extensive consideration for incinerators in the state, given the anticipated challenges of siting and properly operating landfills. As examined in chapter 2, the curiously named "Maximum Incineration Strategy" proposed a network of conventional incinerators across the state: "This strategy assumes that landfill will eventually become an obsolete disposal method in all but the State's most rural areas, because of increasing development pressures and land costs in the growing urban and suburban counties . . . Part of the rationale is that, if incineration will ultimately become the preferred processing method in a given area, the [state] Plan should anticipate this and investments in these facilities should be made at the outset."[92]

The Maximum Incineration Strategy—comprised entirely of conventional incinerators—never made it past the drawing board. But the reinvention of incineration as WTE pushed the technology back to the front of minds of many policymakers, especially in the aftershocks of the 1970s oil crises. Suddenly, the technology was being studied by everyone from the federal Department of Energy and the U.S. Environmental Protection Agency to city governments and local public works departments. The possibility that our nation's trash could offset even some portion of the rapidly growing demand for electricity while also addressing rising volumes of waste seemed like a win-win.[93]

It appears that the state of New Jersey's first official study of WTE was actually in the context of developing the state's Energy Master Plan in the mid-1970s.[94] Observing that "Continuing the present solid waste management system of landfilling valuable raw materials has become an increasing concern because of the problems of our nation's balance of payments in international trade and dependence on foreign crude oil," the authors of the study from the State Department of Energy proceeded to make a case for WTE in New Jersey.[95] The unnamed authors of the report argued that "energy concerns could be a mechanism to elevate [solid waste] from the shadows of the past and into the limelight as an opportunity for meeting our important energy goals . . . with the hopes that the energy potential that exists in our solid waste can finally be harnessed for productive use in the State."[96] The report claimed that investing in WTE would be a boon for New Jersey's environment and economy alike. The authors estimated that were the entirety of New Jersey's waste stream treated using WTE, that the annual electrical energy needs for over 800,000 New Jersey homes could be met. In particular, the report suggested pairing WTE with recycling, to recover salable metals and newsprint. From the department's perspective, recycling represented an energy savings since new materials did not need to be produced. Taken together, the authors estimated that a combination of WTE and intensive recovery of metals and newsprint could yield the electrical energy equivalent of over 60 percent of the residential housing in New Jersey.[97]

Some counties in New Jersey had already been investigating an integration of WTE with recycling as a way to meet their state planning obligations and cut a path out of the waste disposal crisis. By 1983, Essex County was in the process of updating its comprehensive waste management plan to reflect the embrace of both strategies. County officials cited the passage of the 1981 Recycling Act—and its target of at least a 35 percent recycling rate—as all but requiring the reevaluation of the county's earlier commitment to WTE: "Given the necessity of committing a specified amount of solid waste to the energy recovery project . . . a concern developed that recycling options would be foreclosed if the role of recycling in the overall solid waste management system was not clearly delineated."[98] In other words, the county feared that any new recycling mandate would alter the flow of material to their proposed WTE facility. Since WTE facilities are designed, financed, and operated based on their daily capacity, a reduction in the amount of material flowing through the furnaces would negatively impact both operating efficiency and economics. At the same time,

Essex County noted that the benefits of integrating recycling and WTE would most likely outweigh the costs:

> In addition to concerns about potential negative effects of [WTE] develop-
> ments on [recycling] programs, the County was also aware of numerous
> potential benefits to WTE operations that might result from major material
> recovery programs. There has been a widely held but untested opinion that
> comprehensive recycling programs could directly benefit energy recovery
> plants and improve the solid waste management system. These expected
> benefits from recycling include the following elements:
> - Lower total capital, debt and operational costs as a result of a
>   significant reduction in [WTE facility] size;
> - Substantial savings to municipalities in the form of avoided disposal
>   costs;
> - Increased number of employment opportunities related to the overall
>   county solid waste management program;
> - Strengthened local economy through supplying industries in NJ with
>   recycled raw materials from local sources;
> - Reduced environmental impact.[99]

The notion that the tandem of WTE and recycling represented the best path forward for New Jersey had by the mid-1980s become the official stance of the NJDEP's Division of Waste Management. A 1984 public information pamphlet hailed the one-two combination of recycling and WTE as the "Solution to New Jersey's Garbage Dilemma."[100] After reminding readers about the impending disposal crisis and severe shortfall in landfill capacity, Division of Waste Management authors explained that "Recycling and resource recovery are compatible and work best when used together. Although recycling cannot deal with the entire waste stream it will remove materials of economic value and result in a more favorable and consistent waste product for efficient combustion . . . Even with statewide mandatory recycling and a high degree of compliance, however, there will still be at least 55–60% of the total waste stream remaining for [WTE disposal]."[101]

By the time of publication of an update to the State Solid Waste Management Plan in 1985, NJDEP outlined "three basic concepts" in its guidance to counties and municipalities regarding planning for disposal moving forward:

> First, the Solid Waste Management Act provides explicit guidance, mandating
> district waste disposal strategies to "include the maximum practicable use of
> resource recovery." In implementing this guidance, the Department has
> required districts to plan for the implementation of resource recovery facilities
> [WTE], in the medium-to-long term.

The second basic concept is that both landfills and resource recovery facilities are necessary and must be constructed and operated according to state-of-the-art standards. An important corollary is that districts should be extremely careful, in selecting resource recovery technologies, to avoid untried and unreliable technologies.

Finally, the Department recognizes the important role recycling must play in a balanced waste management strategy, and has urged all districts to develop this option to the fullest extent possible.[102]

In other words, by 1985 NJDEP policy was for each New Jersey county to pursue development of a WTE facility, and one that would work in collaboration with emerging recycling schemes. The NJDEP made clear that it would not approve revisions to county plans, or really any other county-level actions, unless and until counties could demonstrate their pursuit of WTE. Very shortly, the message was received by New Jersey's counties: some ten WTE facilities had already been sited by 1985 with several others in various stages of the planning process.[103]

While the logic behind an intensive deployment of both WTE and recycling was attractive to many regulators and waste management experts, considerable doubt regarding the "maximum practicable use" of WTE lingered among elected officials and segments of the broader public. Perhaps first and foremost among the concerns was air pollution. Old-style incinerators had left a terrible legacy in the minds of many, prompting an understandable knee-jerk reaction against any proposal that looked to build a waste management strategy that relied on burning trash. Additionally, as plans contemplating a network of WTE facilities in New Jersey began to emerge, many were aghast at the estimated costs. These concerns are well illustrated in the 1984 testimony of Assemblyman Frank Pelly, who offered the following observations in a hearing concerned with expanding the tax on landfilling wastes to also generate revenue to finance construction of new WTE facilities:

> Our State is rapidly reaching its landfill capacity. It has the obligation to address this issue, and to address it in a very aggressive manner . . . the construction and initial operation of resource recovery facilities—as we are all finding out in our respective districts—are highly capital-intensive.
>
> As I was coming here this morning, I heard a piece of news on our local radio station, expressing the fact that over $1 billion will be spent on resource recovery in the State of New Jersey when it is finally brought to fruition.
>
> As we proceed in the area of resource recovery, we have problems . . . Those problems are coming from the people in general who are concerned with the siting, the construction, and the stack emissions that are potentially going to be coming out of the resource recovery plants . . . we are going to continue, both individually and collectively, to struggle with these issues. We are going to

struggle with the questions of: Where are we going to put resource recovery plants? How are they going to be constructed; and who is going to monitor acceptable air emission standards? In fact, what are acceptable air emission standards?

Hardly a week goes by that one group or another, in one county or another, does not express its fears and apprehensions about the location of these plants. The fears and apprehensions are real, and they are genuine.[104]

Thus, almost simultaneous with adoption of the NJDEP's position that a tandem of WTE plants and intensive recycling efforts could be the clearest and most decisive pathway out of the state's solid waste management crisis, was the political realization, among many of New Jersey's elected officials at municipal, county, and state levels that actually implementing the plan—*selling* the public on building incinerators in their communities and shouldering a potentially high cost to do so—represented a form of career suicide.

Environmental groups seized on the twin specters of air pollution and financial ruin to pitch even more intensive recycling as the best solution to the waste crisis. For instance, the Environmental Defense Fund published in late 1985 a studying examining "the economic advantages of recycling over garbage incineration."[105] Focused on New York City, which floated a plan, similar to New Jersey's, to build a WTE plant in each of the five boroughs, the Environmental Defense Fund's authors observed that recycling "can have significant advantages over incineration . . . in strictly economic terms," and that "environmental and health considerations . . . represent an important advantage for recycling as compared to incineration."[106] The authors continued, "Obviously, a recycling program will not only avoid the air pollution that would come from a like amount of incineration, but will also avoid the environmental impacts associated with obtaining a like volume of virgin resource (for example, making it unnecessary to cut down as many trees, or mine and smelt as much metal, or quarry as much gravel, or scrape off as much topsoil—as well as avoiding associated energy consumption."[107] Dozens of hearings were held and studies were commissioned to examine the potential impacts of WTE in New Jersey; not in the abstract, but rather in instances of particular sites, technologies, and locations.[108] Many of these hearings were heated, contentious, and borderline violent, requiring police protection for the public officials present.

The combination of strong public opposition and limited willingness among most elected officials to work through siting and development processes resulted in little progress for the tandem of WTE and recycling infrastructure that was once hailed as the "Solution to New Jersey's Garbage Dilemma." As debate about whether WTE could be the solution New Jersey was looking for raged on in the late 1980s, real trash continued to pile up and neither debates about WTE nor attempts to enhance the recycling rate resolved the immediate crisis of rapidly dwindling landfill disposal capacity. Many counties and waste haulers quietly shipped waste to landfills in Pennsylvania, New York, and elsewhere, hoping to avoid a calamity as new facilities were debated and planned.

In 1987, legislation was considered to grant the governor and the NJDEP commissioner additional powers to impose directives on counties that were "certified failures" in the waste management planning process—those who could not demonstrate any progress toward additional disposal capacity.[109] The commissioner would have been granted the power to impose sharp financial penalties on noncompliant counties, and also to impose (or override) any county plans regarding "waste flow" of trash to particular disposal sites, enter into new contracts on behalf of the county, purchase land for disposal sites, and issue emergency authorizations and approvals. The bill's sponsor, Assemblyman Bob Shinn, observed during a hearing to examine the legislation, that "We have essentially two major facilities in the northern part of the State that are receiving about 60% of the State's volume of trash, those being Edgeboro [Middlesex County] and HMDC [the Hackensack Meadowlands area] . . . I guess the bottom line is, the State is closing landfills rapidly, and is running out of capacity very rapidly. So there is a need for implementation of all of the counties' solid waste plans and, in lieu of that, a mechanism to provide either capacity on a short-term basis or transfer stations to get trash to the appropriate facilities. If those facilities are not going alone, we are, indeed, entering a state of crisis within 12 months."[110] Counties that would be immediately impacted by the new laws, like Morris County, responded that it had been the NJDEP that had failed the citizens and businesses of New Jersey, not the counties themselves. The solid waste coordinator for Morris County, Glenn Schweizer, responded that

> The Bill continuously refers to failures upon failures by the counties. We feel that DEP and the Legislators should delete the word failure as it applies to counties unless and until they are able to address and correct their own failures . . . Morris County has attempted in the past to work with the State to solve our solid waste problem and we were unable to succeed due to federal intervention. Passage of the Superfund Bill precluded us from developing a landfill at the approved site.
>
> And now, the County is being selected for punishment. In fact, counties and municipalities are already being financially punished through importation taxes and high transportation costs. The proposed penalties are both confusing and punitive . . . [Waste management] goals should be addressed on a mutual, cooperative basis and not through antagonistic and punitive measures.
>
> This bill sends a bad signal to any cooperative relationship between counties and the State, especially through the proposed punitive assessments.[111]

While the legal wrangling may have remained somewhat obscure, a spectacular meltdown of the United States' waste management policies and infrastructures would capture the attention of the public, in New Jersey and around the United States. The *Mobro 4000* barge was chartered to haul waste from Long Island, New York, to a disposal facility in North Carolina. However, the ship was turned away from its original destination, spurring a months-long voyage along the East Coast and Gulf Coast of the United States and even into the Caribbean as it sought a place to unload. Eventually the *Mobro* returned to New York, where its cargo was burned in an incinerator in Brooklyn. Though fears that the barge contained toxic waste, hospital waste, or even a deceased contingent of New York City's rat population proved to be unfounded, the bizarre story of a homeless "gar-barge" prompted greater scrutiny of all components of the United States' waste infrastructures.

In many ways, the *Mobro* incident was a major win for proponents of recycling. As it emerged that a considerable volume of the material on the *Mobro* was paper fiber of one type or another, environmental activism group Greenpeace attached a banner to the barge reading "Next time . . . Try Recycling!" While in New Jersey, considerable time and energy was already being devoted to precisely that notion, the idea that recycling offered immediate relief to the waste disposal problem found considerable traction. Despite the substantial concerns of many in the NJDEP, manufacturers who would ostensibly purchase recovered materials, and even those in the recycling processing industry that markets for many materials were weak—or in the case of the rapidly growing category of plastic materials, entirely nonexistent—improving the recycling rate was moving toward becoming the preferred waste management strategy for the state. Accordingly, in the 1989 gubernatorial election, both parties backed away from a statewide network of WTE plants. The state's Republican Party (the party of outgoing Governor Tom Kean who had supported and publicly advocated for the WTE and recycling tandem) said in their 1989 party platform that "Alternatives to incineration of solid wastes must be explored and innovations in the biodegradability of packaging materials should be examined," before noting that "the mountains of trash, however, cannot wait for new technologies." The party platform continued, "Recycling has begun with a goal of reducing the waste stream by twenty-five percent. Plastics will be the next great recycling frontier and will be part of the state's program for dealing with solid wastes. Efforts should be undertaken to encourage new markets and new uses for recyclables. The state should lead efforts to use recyclable materials . . . We support efforts to educate and inform citizens about the dimensions of waste disposal programs and the facts concerning disposal alternatives."[112]

Democratic gubernatorial candidate, and eventual winner James Florio, was even more direct in his opposition to WTE. As discussed at the start of this chapter, Florio's platform specifically halted most of the WTE facility planning process and called for increased investment into recycling. Upon taking office, Governor Florio initiated a complete review of New Jersey's waste management infrastructure along with the state's overall approach to waste management planning that resulted in much different outcomes from what proponents of either WTE or recycling had been striving for during the late 1970s and 1980s.

5

Limits to the System

• •

> New Jersey is not a separate planet.
> —Albert A. Fiore, Executive Director,
> Hudson County Improvement
> Authority, May 31, 1989

Shortly after taking office in 1990, Governor James Florio issued an executive order establishing a fifteen-member Emergency Solid Waste Assessment Task Force. In the order commissioning the task force, the governor laid out a vision for the future of waste management in New Jersey, nearly opposite from the one that had been championed during the 1980s:

> WHEREAS, the lack of coordinated Statewide planning and management has led to insufficient disposal capacity within the State . . . at great cost and questionable reliability . . .
>
> WHEREAS, source reduction, reuse, recycling and composting efforts reduce demand for solid waste disposal facilities and, conversely, waste-to-energy resource recovery facilities discourage the maximum use of other recycling activities . . .
>
> WHEREAS, a sensible plan for dealing with the State's solid waste problem can be achieved by maximizing the use of source reduction and reuse techniques, recycling, composting and other environmentally sound methods for dealing with solid waste, by reassessing options for landfilling, and by reassessing those waste-to-energy facilities that are currently being developed or are operating in this State.[1]

Florio had run, in part, on a pledge to halt the development of waste-to-energy ("WTE") facilities across New Jersey and pursue instead higher levels of recycling, even while many stakeholders across the universe of waste management in the state wondered very openly about the economics of recycling, particularly plastics. Many materials could be collected, they acknowledged, keeping them from the state's landfills—but without purchasers of recyclables, what was the real benefit?

At the same time, even prior to Florio's election, policymakers in New Jersey were having second thoughts about the network of waste-to-energy (WTE) incineration facilities that had previously been proposed as the *solution* to the state's disposal capacity woes. WTE was expensive and despite being a vastly improved technology, conjured up images of dirty, inefficient incinerators of years past. Public hearings relating to the siting and operations of virtually any WTE facility became contentious, and even violent. In light of their constituents' concerns about WTE, many elected officials in New Jersey found it easy to oppose the technology even if they had previously been favoring it.

Thus by 1993 and the publication of a new statewide plan for solid waste management, the push for a network of WTE facilities in New Jersey was all but dead. Instead, the state had committed to much more intensive plans for recycling than what had been initially imagined in the 1980s. Furthermore, amid the ramping up of recycling in the 1990s, and alongside the operation of new WTE and sanitary landfill facilities that had been planned and permitted during the "crisis" years of the 1980s, the pressure for new disposal capacity in New Jersey eased somewhat. A shift in strategy was feasible.

However, if the previous decades had been focused on building up New Jersey's waste management infrastructure through an intensive process of planning, rulemaking, financing, and public education, the 1990s seemed dedicated to unraveling much of that work. The remainder of the twentieth century and early years of the twenty-first saw the sudden end of waste "flow control" and the removal of many of the supports for financing New Jersey's waste infrastructure, especially—and ironically—recycling. A great deal of the innovation that had been pursued earlier was put aside, as policymakers were forced to reconsider the entire system for collecting and disposing of waste in the state, even as new and even more challenging types of wastes emerged, mounting piles of public debt loomed, and natural disasters like Superstorm Sandy created tsunamis of trash.

This chapter reviews in greater detail the decision to back away from an extensive network of WTE facilities in New Jersey, a concept which had been the cornerstone of waste management planning in the state for most of the 1980s. Readers will recall that the WTE network, in conjunction with comprehensive recycling, was the consensus solution to New Jersey's waste disposal woes. In place of WTE, policymakers decided to go all-in on the so called crap shoot of recycling, hoping that aggressive diversion of a broadening menu of materials could be coupled with (wishful) cajoling markets into existence for those materials. It is important for readers to understand that since the early 1990s when this decision was taken reliable markets for most recyclable materials and specifically plastics have yet to be created, meaning that a central pillar of New Jersey's recycling-first plan for waste management has never really come to fruition. The chapter then moves to examine the court-ordered dismantling of another pillar of the state's waste management strategy, so called flow control ensuring disposal of wastes at officially designated facilities. The *Carbone*, *Atlantic Coast*, and related legal decisions removed vital financial support for the disposal facilities still remaining in New Jersey as haulers would be free to choose any disposal site they would like regardless of state or county directives. These legal outcomes also hamstrung the finances of New Jersey counties that had built new disposal facilities with massive "stranded" assets, as revenue from waste fell sharply. The end of flow control became an all-consuming matter for waste management stakeholders in New Jersey at the end of the twentieth century, even as the new millennium's challenges started to take form.

Out with WTE, in with Recycling

Shortly before the 1989 gubernatorial election, the Energy and Environment Committee of the state senate held a hearing to consider the "appropriate role of incineration, and its alternatives" in New Jersey's waste management system. The chair of the committee, state senator Daniel J. Dalton, observed in his introductory remarks that "solid waste incineration is one of the most significant environmental issues facing our State . . . because of the complex environmental and public health issues associated with properly burning massive amounts of solid waste . . . Once the financial commitments have been made to these incinerators, we will be ultimately dependent upon them, and we will have no choice but to use them."[2] Dalton and the other state senators present— Catherine A. Costa, John D'Amico, and William L. Gormley—grilled those appearing before the committee about the environmental performance of WTE, the costs, and especially, the relationship between WTE and recycling. Others present at the hearing, including representatives of environmental advocacy groups but also the waste management and WTE industry itself, raised any number of challenges to the New Jersey Department of Environmental Protection's (NJDEP) plan for a WTE facility in every county.

For many of those participating in the hearing, it seems that frustration with the NJDEP itself over perceived (and sometimes actual) delays in permitting, enforcement, and other administrative matters mutated into opposition to NJDEP plans in general. Others relayed their resentment of NJDEP's attitudes toward county and local governments, portraying the agency as a steam-roller bent on imposing a vision for self-sufficiency in waste disposal capacity over local opposition and popular out-of-state exporting practices. Further-more, environmentalists alleged that existing WTE and incinerator facili-ties in the state were already operating afoul of environmental standards, and feared that air quality would be impaired even further with each incin-erator that was built.

During the hearing, Christopher J. Daggett, Dr. Donald A. Deieso, and John Czapor—NJDEP commissioner, assistant commissioner, and director of the Division of Solid Waste Management, respectively—spent considerable time facing the senate committee. The trio carefully explained how WTE still represented the best solution for the state's waste disposal needs moving for-ward. Given the limited space available for landfills, WTE was a disposal alternative with "a proven track record . . . available on a commercial scale . . . [featuring] high temperatures that destroy anything [toxic] that's there . . . [and offering] comprehensible environmental consequences."[3] Other technologies were yet unproven, and thus unworthy of investment of the public's money. According to Assistant Commissioner Deieso, processes like composting, 100 percent recycling rates, pyrolysis gasification, and even a scheme proposed for Ocean County involving the use of lasers to destroy trash were "not New Jersey's answer to the crisis, because they are rather immature . . . They are ideas for the future. We should position ourselves to encourage them, but I suggest to you, they are not here today. They are not available today. New Jersey's crisis is higher today."[4]

The NJDEP representatives appearing before the state senate committee acknowledged that a WTE facility in each county, as originally proposed, was likely overkill. They discussed possible scenarios for "regionalization" where counties could share the benefits, costs, and operations of a single large WTE facility as opposed to building multiple smaller ones, while remaining commit-ted to the broader plan for WTE to become the cornerstone of disposal in the state. And at any rate, NJDEP commissioner Daggett argued, after years of study and deliberation about how best to defeat New Jersey's waste disposal crisis, there was finally momentum: "Let me say, what we're really trying to say is, this isn't a time to really take pause and maybe stop, or go for a moratorium. The fact is, we think we've got to continue to move forward on this process. Indeed, we may have to make a few in-flight corrections. We have to press for things like regionalization. We have a waste strategy in place. I think it's been effective . . . We're moving forward for the first time in some time. I think we need to continue that, and I think that we can only do that by moving forward."[5] None of these arguments would be enough. Concerns about potential air emissions problems, huge cost overruns and a contentious siting process were already cemented in the minds of many even though NJDEP's research suggested otherwise.

But, the core of the opposition to NJDEP's WTE plan turned on the optimism that recycling *could* be improved in the future: active markets *could* be developed, and uses for troublesome materials like plastics *could* emerge. These possibilities suggested that investments in large WTE plants would be foolish because as the recycling rate increased, the WTE facilities would be immediately underutilized. Furthermore, recycling looked comparatively inexpensive when compared with WTE:

ASSISTANT COMMISSIONER DEIESO: Would recycling then be the cheapest if markets were available?

SENATOR D'AMICO: Or cheaper?

ASSISTANT COMMISSIONER DEIESO: Cheaper, yes, but we need to pull back from an excursion into the future and say, are markets available, and what prevents them from becoming available? It isn't an on and off switch, Senator. Let's recycle plastics, and this plastic today—That market development is one that is going to take years to establish. Some recycled materials are not going to be cheaper to recycle . . .

I think it's one of these areas where we say to you, "We' re all for recycling." We encouraged it, and have the first program in the country. But, to go from 25% to 50%, is more than just the doubling of a number. It means that we're going to have to create the markets and the separation facilities. We're going to have to recycle the materials that heretofore haven't been thought of . . .

COMMISSIONER DAGGETT: But, it's not something that's going to happen in the next year or two, by the time you develop markets, change social behavior, and do all the things we have to do.[6]

The conversation about further enhancing recycling rates quickly became heated.

SENATOR DALTON: So what you have to do, it seems to me, is, you have to force the technology. Your whole concept of crisis, the imminent peril—okay— You're saying that incineration is the answer to this imminent peril, to move forward with incineration. My feeling is, "That ain't enough." You better be moving forward with recycling as well.

COMMISSIONER DAGGETT: There's also the question of—and I think we're certainly willing to enter this discussion and try to reach some sort of—

SENATOR DALTON [INTERRUPTING]: Don't give me phrases like "enter this
discussion." You have to be aggressive.

COMMISSIONER DAGGETT: Of course you do, and we're very aggressive. I don't
want it to be portrayed that we're not. I think we are very aggressive in
these issues. The question is, how long do you want to play, essentially, a
crap shoot on this? We don't know that [recycling facilities] will be able to
be sized up and handle any kind of a volume. We do know, if we make the
commitment, we can build incinerators within a certain time frame and be
able to handle that . . .

SENATOR DALTON: The whole thing is a crap shoot, okay.

COMMISSIONER DAGGETT: Exactly right. How long do we want to play that
out? On the short term, we can put in facilities. In the longer term, we can
put in the ultimate technologies. The question is, do we want to go for the
longer term or not? That is an open question . . . In the interim, do we want
to wait or don't we?[7]

While elected officials were questioning the logic of the state's WTE-
oriented plan, the potential success of recycling appeared to know no limits.
In 1990 the state released its first report on recycling since the introduction of
the mandatory recycling law a few years prior. *Recycling into the '90s*, released
by the NJDEP's Office of Recycling, asserted that: "Since 1987 recycling has
become a way of life for New Jersey residents. In 1988, 2,749,345 tons of material were recycling in New Jersey, as reported to, and documented by, the Office
of Recycling. This is up from 1,830,000 tons recycled in 1987. This equates to
24 percent of the total waste stream which includes paper, metal, glass, plastic
containers, yard waste, auto scrap, concrete, asphalt and wood waste. 1989 figures should document that New Jersey has surpassed its mandated goal [of
25 percent recycling rate]."[8] The report, and others like it, cast recycling as an
engine of innovation in the state, spurring the growth of new businesses to
handle these materials, investment in existing firms, and even growth in corners of municipal and county government, all equating to new jobs and opportunities. Furthermore, all of these benefits came at an apparently low cost. The
1987 Recycling Act applied a $1.50 tax per ton of waste disposed at the state's
landfills and incinerators, generating about $12 million for the recycling fund.
In turn this fund supported efforts for municipalities and counties to expand
collection services, provided a pool of capital to lend to recycling-oriented businesses, and create public outreach materials aimed at both adults and children
(through a school curriculum development program).[9] The tax, however, was
not a permanent source of revenue for recycling—it needed to be extended
through an act of the legislature every few years. Yet, the data and projections
for recycling tonnages included in the report showed a steady increase in the

amounts of material recovered each year, from under 1 million tons recycled in 1985 to nearly 3 million tons recycled in 1990.[10] Compared with the siting, construction, insurance, pollution control, and other costs associated with building even a single WTE facility—reaching into the tens of millions of dollars each, if not more—recycling appeared to be an absolute bargain.

Given the rapid success of the state's recycling efforts, and comparatively low costs, it is easier to understand why so many were able to look past recycling's glaring limitations. Furthermore, New Jersey was by 1990 a national leader in designing and implementing recycling systems and devising policy solutions to support those systems at municipal, county, and private industry levels. But to shift the public mindset about recycling from being a complementary activity to being a primary strategy for managing New Jersey's waste volumes would require a conscious and consistent effort reaching beyond the state's households and dwellings. In a hearing conducted shortly after the release of *Recycling into the '90s*, John V. Czapor, director of the Division of Solid Waste Management within the NJDEP once again appeared before policymakers, this time the state Assembly Waste Management, Planning, and Recycling Committee, to make such a case. Czapor argued that "It is appropriate, I think, as we move forward from where we are now into increasing recycling, to change our focus and our philosophy a little bit from recycling as a means of conserving resources and environmentally sound policy, to recycling as a solid waste technique that reduces our dependency on disposal facilities ... In other words, dealing with the whole waste stream, regardless of whether it is municipal, commercial, or industrial; regardless of whether it is yard waste or construction and demolition debris. All of these take up landfill space if not reused, recycled, or reduced, all of which would use up our precious land resources."[11] Czapor spoke candidly about what the shift in policy would entail. Previously, the concept of recycling had traded on the notion that it offset its own costs through the resale of the materials recovered. But to choose recycling as a major waste *disposal* strategy would require acknowledgment that certain recycling activities—dealing with plastic, for example—would likely operate at a loss. Czapor noted to the committee:

> I think, in conclusion, from my perspective, there are two aspects to recycling that need to be focused on. One is basically where it is just sound environmental policy to recycle from the perspective of conservation of resources and just good environmental ethic; and [the other is] where we start to incur costs—costs to the individual, costs to the municipalities, costs to the counties, where recycling is being done as a solid waste management technique ... What do we achieve in terms of reduction of the material that needs to be handled?
>
> I think we can take each issue, each component of the waste stream, and really put it to the acid test of how much [time and landfill space] it will buy you in terms of targeting increased recycling [of that material].[12]

The state assemblymen and assemblywomen present at the meeting, though unanimous in their skepticism of the NJDEP's plan to build incinerators in each of New Jersey's counties, were clearly apprehensive about the potential for increased recycling to solve the state's waste management problems. Assembly-man Bob Shinn, who himself played a leading role in the development of Bur-lington County's solid waste management infrastructure, had some of the most incisive concerns.

> I guess generally we are promoting a very positive picture of recycling, but I guess my feeling is we have a lot of problems with it . . . I think the thing that is going to drive the numbers in recycling, if we are real honest about the whole picture, is the dollars we are going to get for material . . . There is an awful lot of materials in that waste stream that still have to come out. How you get them in a practical sense and find markets for them is the real issue. How do we develop these markets?
>
> I just see the lion's share of the burden keeps—We seem to be saying, "Okay, municipalities and counties, you go do it." There are limited markets out there at best, with a high subsidy factor, and I know we have a good recycling grant program—limited, but good—but we still ship the lion's share of this recycling subsidy, if you will, to municipalities and counties. We just sort of have them in the middle. Without that market loosening up—You know, something has to give. I guess that is my point.
>
> I like the general tone of what you [Czapor] have put before us, but there are a lot of concerns I have about really raising these levels. The worst thing we can do is lull ourselves into believing we can recycle our way out of the solid waste crisis, without other serious alternatives and large percentages to deal with the waste stream.[13]

The "plastics question" loomed large—what could be done with these increasingly ubiquitous materials, and who would purchase them? Interestingly, at the hearing representatives from several different corners of the plastics and polymers industry were present, along with representatives from newsprint and glass producers (who were also involved in the recycling of newsprint and glass as well). In response to concerns about the markets for plastics, industry representative assured the members of the committee that there was in fact a current *shortage* of plastic available to even be recycled. That is to say, the plastics industry claimed it could not source enough recovered plastic from recycling programs to actually make new products out of recycled plastic.

However, whereas glass and newsprint processors were able to quote recent prices for both recovered glass and newsprint as well as remanufactured products, plastics industry representatives were unable to cite any figures. Whereas newsprint and glass were bought and sold in a market with prices changing based on supply and demand, recovered plastics—to the extent that they were traded at all—were exchanged privately between buyer and seller.

It is clear from the testimony that representatives from the Plastic Recycling Corporation of New Jersey, James River Corporation, Union Carbide, Polysource Mid-Atlantic, Day Products, Sonoco Graham, Wellman Company, Wheaton Plastic, and Pepsi Cola—all major plastic producers and consumers—were angling to have plastics added to the list of materials that New Jersey's towns and counties were obliged to collect through their recycling programs. However, the plastics industry was loath to have to purchase these recovered plastics in a transparent market—preferring instead to pursue *exclusive* contracts between their processing facilities and individual towns, counties, or recycling processors. That way, quantities and terms could be hidden from competitors, and the plastics firms could also exert monopoly-like dominance over the prices sellers would receive. While this would surely represent a dedicated end use for the recovered plastics, should the plastics firm move its operations, go out of business, or simply require a different type of material for its operations, there would all of a sudden be scads of plastic being collected with no buyer in sight.

This scenario was one of the reasons why a commitment to market-based pricing, which was initially intended to encourage towns and counties to recover only the materials which had a high likelihood of being resold, was baked into New Jersey's recycling statutes. This was one of the major ways that recycling could be sold as a lower-cost solution to the waste disposal problem. But by seeking to leverage New Jersey's emerging recycling infrastructure as a way to secure low-cost supplies of materials to their own manufacturing activities, the plastics industry was avoiding any market-demand based mechanism entirely. This was troubling to some of the policymakers present at the hearing. As an exchange between Assemblyman Shinn and James B. Rouse, a regional director of public affairs at Union Carbide illustrates:

> MR. ROUSE: . . . the barriers facing our recycling project are really human, not technical. We know that plastics, of course, can be recycled. We have the technology, the skills, and, more critically, the end-use markets for recycled products. The big problem is the lack of recyclable material. Until recently, too few communities have been including plastics in their curbside collection programs, but New Jersey is leading the way in the Northeast with nine of 21 counties . . . that are including PET and HDPE [two types of plastic resins] in their curbside programs.
>
> State and local governments . . . have been reluctant to commit to plastics collection and recycling. There has been really very little reason to take the risk. Many are concerned that collection methods are cumbersome and require reconfigured bins and garbage trucks. Others think there

are no end-use markets for recycled products. Well, I can assure you, Mr. Vice Chairman and other members of the Committee, that there are end-use markets.

... [We] need for local officials to institute plastics recycling to ensure that there is more than lip service to plastics recycling. Union Carbide will be part of the growing national partnership of private sector and public interest in working toward solid waste solutions.

ASSEMBLYMAN SHINN: This is the question, just a little dichotomy in some of the philosophy you are recommending. You are saying include it as a mandated recycling item, and tell the counties and the towns to recycle plastics. I think the general thrust of the legislation was, let the market drive what is most cost-effective to take out of the waste stream. I guess therein lies the point of contention. Once we direct it to a private corporation, we take out the cost-building process of what you are going to pay for recycled materials because it is mandated at your doorstep.

The problem I have in this whole issue is, if we mandate municipalities and counties to do a specific item, then we sort of downgrade the marketability of that from the manufacturer's standpoint.

MR. ROUSE: What is different now—

ASSEMBLYMAN SHINN [INTERRUPTING]: You are not talking market prices, though. See, I am talking market prices. You are talking, "Let's go recycle—rah, rah, rah." I'm all for that, but let's talk about what kind of market prices are out there and what the range of market prices [for plastics] has been ... Where did we start, and where are we now and what does the future hold?

MR. ROUSE: Well, the market—I can't get into specifics, but I can tell you that the market price right now that we are paying is tied to the price of virgin polyethylene. So the market price for recycled materials, or raw materials, if you will—the plastic bottles—will vary according to the market for the virgin material. Not only that, but it will vary according to whether [competing firms] are out there trying to enter into agreements. That will drive the price up.

ASSEMBLYMAN SHINN: Yeah, but you are still not telling me anything.

MR. ROUSE: I can't give you any market guarantees as to what—

ASSEMBLYMAN SHINN [INTERRUPTING]: Yeah, but you are not giving me anything. You are not telling me a range, a history, or—

MR. ROUSE [INTERRUPTING]: Well, we have—

ASSEMBLYMAN SHINN [INTERRUPTING]: I'm sure the Legislature is going to look, before they direct a stream to a corporation to see that there is some equity balance. Are we suppressing that market and suppressing a revenue that is going to come out of that market? Are we encouraging a free material supply?

Now, depending on what you can use the recycled material for—if it is the same price as raw material, the raw material is going to drive the recycled material up, unless we give you a free lunch by saying, "We are going to direct everybody to send their plastics to you." How do we keep that at an honest market fluctuation?

MR. ROUSE: You will never direct anybody to send it to us. You are talking about us as an industry, us as a business, not us as Union Carbide . . . Now, as long as the [manufacturers] specify a certain amount of recycled content and we have to get that recycled resin, we are going to be going out there and looking for plastics . . . it was the chicken and the egg . . . Who is going to make the first move? Well, the plastics industry has already made the first move, and we now have results . . .

ASSEMBLYMAN SHINN: I think the first move was the Mandatory Recycling Act. It sort of drove the plastics industry into action . . . Then the plastics industry sort of responded and is getting geared up to address this market. The next move, to me, is pretty critical; There has to be a good balance in what we do in the next step. I think that is significant . . . to give this this Committee some assurance that there is going to be a strong market for plastics and make everyone feel better in moving forward . . .

MR. ROUSE [INTERRUPTING]: I can't give you that, because, quite frankly, I am really not sure what the price range is.[14]

Even as concerns about the relationship between recyclable materials and markets for their purchase remained unresolved, the shift in strategy away from WTE and toward enhanced recycling was underway. Early in his tenure as governor, Florio announced a moratorium on the construction of new WTE facilities, effectively forcing the state to reevaluate its entire waste disposal strategy. Midway through his first year in office, Governor Florio's Emergency Solid Waste Assessment Task Force issued its findings, which the governor subsequently accepted, setting in motion a dramatic shift in policy for New Jersey. First and foremost among these was the assertion that:

A solid waste plan for New Jersey must be built around realistic targets for reuse, recycling, and composting. Environmentally and economically, these methods to deal with solid waste provide enormous advantages over other options. Only when the amount of waste that can be eliminated or recycled is calculated can the rest of a solid waste management plan be instituted . . .

State and county solid waste management plans shall be modified to demonstrate that disposal capacity and the management of solid waste in New Jersey will be developed based on the attainment of a 60% recycling rate. The policy of encouraging incinerators as a part of almost every county's solid waste management plan is rejected.[15]

The task force was proposing that the state shift its policies on waste management to focus almost exclusively on recycling, asserting that new disposal capacity (landfills and WTE facilities) would only be approved if counties could demonstrate substantial amounts of waste still needing disposal after the 60 percent recycling rate had been achieved. Implicit in this policy was the threat that towns and counties would be flooded with their own wastes unless and until recycling targets were hit.

Only a few years earlier, the NJDEP had anointed WTE as the cornerstone of waste disposal for New Jersey. The task force acknowledge that its findings represented a "substantial change in direction for solid waste policy in the State."[16] New Jerseyans had only recently achieved the 25 percent recycling rate set out in the Recycling Act, and the new recommendation was aiming to increase the amount recycled by 35 percent, to a goal of recycling 60 percent of the total waste volume of the state, by 1995. To be sure, this included expanding the definition of *recycling* beyond materials like newsprint and aluminum, to include things like tires; concrete, wood, and other construction and demolition debris; white goods; junked automobiles; and additional types of organic wastes (food and yard waste). Describing the dramatic expansion of recycling on several occasions as "ambitious, yet realistic and achievable," the task force cited strong support for the 60 percent figure coming from "public officials, environmental organizations, trade associations, and private citizens."[17] Notably absent from the list of supporters, and seemingly excluded from participation on the task force, were representatives of the recycling industry and manufacturers who could potentially rely on recycled materials.

As a litany of hearings and public testimony in the 1980s had demonstrated, the success of recycling programs in New Jersey hinged on the presence of an active, reliable market for recovered materials. But the task force offered scant few recommendations in this arena:

Recommendation: Strategies must be developed to increase markets for recyclable commodities.

In evaluating the current status of recycling in New Jersey, the Task Force has noted the vital importance of developing reliable markets for recyclable commodities.

Policies are needed to ensure that our State agencies have the specialized capabilities needed to promote the development of markets for recyclable commodities.

. . . The State must act forcefully to expand in-state capacity for remanufacturing and recycling commodities generated by community collection programs

. . . Policies and programs must be developed to create new recycling markets for materials such as paper and construction and demolition waste.[18]

The most substantial specific suggestions focused on mandating that government adopt procurement policies favoring products with recycled content. The task force also recommended a greater role for the state in centralizing and publicizing data about recycling and different materials, a vitally important aspect of any functioning market. For instance, the report suggested that the state provide "timely, accurate market statistics [which] are essential in supporting the planning activities of commodity producers and recyclers . . . standardization of commodity grades . . . collection and dissemination of periodic market statistics; collection and analysis of data on the adequacy of remanufacturing capacity and end markets for recyclable commodities; and collection of market data covering local, Statewide, northeastern US and export markets."[19]

Other suggestions for improving the market for recyclables were less specific, such as "setting recycling targets for individual materials [which] should motivate private industry to absorb greater quantities of recyclable commodities and products"[20] or vaguely defined incentives and financing programs for counties and municipalities. In reviewing the report, it is clear that the task force envisioned the state as the ultimate backstop for recycling efforts, no matter their cost and apparently, no matter how poor the market for a given material. For instance, the task force suggested, without additional context or explanation, establishing "phased-in mandatory standards on the purchase of recycled products . . . to bolster certain lagging markets for recyclable commodities."[21] The task force articulated that recycling services ought to be operated at a loss in order to achieve the 60 percent rate (though this was couched in environmental terms): "It should be recognized that even if the cost of recycling exceeds the cost of disposal, recycling still is preferable to landfilling or incineration due to its clear environmental and land and energy conservation advantages."[22] Certainly, the task force also suggested a few ideas that were perhaps ahead of their time, including extended producer responsibility ("this requirement would place the burden of developing technologies for recycling . . . on the manufacturers"[23]) and imposing requirements for reduced product packaging. The task force suggested a bottle deposit–type program for all manner of packaging and containers, along with regulations requiring that packaging material for products sold in New Jersey be both recyclable and made of recycled materials.[24]

In sum, the task force envisioned a waste management system for New Jersey that relied on the appearance, if not reality, of robust, properly functioning markets for a range of recyclable materials that would effectively cap the amount of material flowing to the state's landfills and WTE facilities that it had managed to build. Where materials could be profitably recovered and processed, markets would be encouraged and supported; where materials would be unprofitable or extremely challenging to recover and reuse, markets would be *forced* to exist and function through legal requirements and (potentially large) subsidies. In all instances, the task force recommended that government be involved in attracting and financing a range of processing and remanufacturing firms to operate in New Jersey.

Despite the lingering uncertainty about details of the new waste management regime, many aspects of the plan were vigorously implemented. By 1993, the NJDEP was asserting that:

> Despite the greatest population density of any state in the country, limited land availability, the presence of sensitive natural resource areas, and unprecedented urban and industrial development, New Jersey has implemented one of the most aggressive source reduction and recycling programs in the nation, reaching an overall recycling rate of 52% by 1991 and projecting a 60% recycling rate by 1996. In addition, New Jersey has developed, and continues to develop, state-of-the-art disposal facilities including landfills, recycling centers, transfer stations, and incineration facilities. Finally, New Jersey has reduced its export of solid waste to other states to 18% in 1991 and expects to be fully self-sufficient [regarding disposal] within the next seven years.[25]

The 1993 *Solid Waste Management State Plan*, reviewed for this project in its draft form, reports the aggressive strategies the NJDEP undertook to implement the task force's strategies. These included, among others items, policy directives for both source reduction of waste and recycled products purchasing mandates that were adopted by virtually all other state agencies and offices, and creating far-reaching educational and marketing campaigns aiming to inform residential, commercial, and industrial waste generators about new protocols for recycling. The NJDEP did also outlined a set of policies and goals for growing markets for recyclable materials in both the short and long term.

The first was simply to require towns and counties (as well as state agencies themselves and New Jersey's colleges and universities) to collect more materials, removing them from the volumes heading to landfills and incinerators. The Recycling Act had required the collection of newspaper, glass, and aluminum cans; "future focus will be upon broader designation of the following materials: grass clippings, brush, office paper, junk mail, HDPE plastic, white goods, used motor oil, consumer batteries, wood waste and other construction and demolition materials."[26] The state also proposed that all construction and demolition permit applications include estimates of wastes that would be produced by the project and specify the disposal and recycling centers that would be utilized, before the project could begin. While the plan articulated strategies for increasing the supply of recycled materials, policies for boosting demand were not as clear.

In the short term, the NJDEP outlined further procurement strategies for government entities, that emphasized the purchase of products with recycled content. Given the size of government purchasing accounts, such a policy would have made a considerable impact on the market for the universe of products that the state of New Jersey might buy. The plan described commitments to investigating the role of recycled materials in various construction and infrastructure projects (for example, the use of crushed glass as an additive in road construction projects), expedited permitting for recycling facilities, and continuation of the low-interest loan program established in the 1987 Recycling Act. However, little by way of stimulating private firms' demand for recycled material was identified, other than the creation of a "corporate recycled products advisory council" and similar "conferences, seminars, and other outreach activities which will involve the various levels of government and the private sector [for] establishing a recycled products purchasing network which will serve to expand and diversify markets for recycled products."[27]

With the 1993 *Solid Waste Management State Plan*, the push for a network of WTE facilities in New Jersey ended. Indeed, as the facilities—WTE and sanitary landfill[28]—that had been planned and permitted during the "crisis" years of the 1980s came online, and the implementation of more-intensive versions of recycling ramped up, the pressure for new disposal capacity in New Jersey eased somewhat. However, the success of this comprehensive system for waste management in New Jersey would last only a few short years, as two guiding concepts—self-sufficiency in disposal capacity and waste "flow control"— would each crumble under numerous legal challenges.

The End of "Flow Control" in New Jersey, Sort Of

It is no exaggeration to say that for the remainder of the twentieth century, a single issue dominated policymakers' waste management agendas in New Jersey: responding to the legal challenges wiping away both waste "flow control" and self-sufficiency concepts that had underpinned the state's waste management planning concept for more than a decade. As the 1993 *Solid Waste Management State Plan* observes,

New Jersey has controlled the flow of solid waste since ... December 6, 1982 ... Waste flow control is a critical component of New Jersey's solid waste management program. All 567 municipalities in the state are directed for the purpose of disposal to specific facilities pursuant to rules found in NJAC 7:26-6 et seq. Through these rules, the state has the authority to direct the flow of solid waste. This power, when combined with data on the amounts and types of solid waste generated in each town, enables the 21 counties and the state to plan and construct properly sized facilities and ensures that a reliable source of material is available.

In addition, this is essential in terms of satisfying the economic commitments associated with solid waste facilities, since the industry is regulated as a utility and disposal rates are specifically set in consideration of projections of the amount of waste which will be delivered. Waste flow control is important in gaining facility financing since a steady source of waste can be identified and essentially guaranteed for delivery. This provides comfort to financial institutions and bondholders that sufficient project revenues will be generated to pay debt service on and [meet the] operating costs of the facilities.

Waste flow control is applicable to facilities such as transfer stations and landfills, but is not applicable to the movement of recyclables.[29]

The ability of the NJDEP and counties of New Jersey to direct the flow of garbage to particular sites was one of the most foundational elements of the state's entire strategy for waste management. As the passage above explains, "waste flow" or "flow control" facilitated the planning process. In the system that had evolved since the 1970s, state and county officials first determined just how much waste there was, then determined which locations had the capacity to accept the waste. As new landfills and WTE facilities were sited and came online, government entities could point to flow control in the process of getting their financing together. County and local borrowers could guarantee bondholders that they would be able to repay their loans, since flow control assured a regular amount of waste to a facility each year, and thus a regular flow of cash collected via disposal fees. For the same reasons, flow control was a critical component in projecting the need for additional disposal facilities, because planners could estimate approximate dates for when a given landfill would reach capacity. Finally, flow control played a role—if only a theoretical one—in stamping out the operations of organized crime and other types of illicit disposal. Because the state was gathering more and more data each year as to how much waste there *should be* coming from each county, and could match that figure to how much waste was actually being disposed at each county's designated disposal sites, glaring discrepancies might point to waste being diverted elsewhere, whether abandoned lots or late-night, unapproved shipments to landfills and dumping sites in Pennsylvania, New York, or elsewhere.

Beyond these practical considerations, flow control played an important symbolic role in New Jersey's strategy for waste management. After the decision of *City of Philadelphia v. New Jersey* in 1978 struck down the state's rules banning the disposal of wastes coming from out of state, NJDEP and other waste management planning officials gradually moved toward a goal of self-sufficiency in waste disposal—meaning, that all the wastes generated in New Jersey would be disposed of within state borders. Furthermore, "Based upon past experiences where New Jersey counties were cut off from the use of specific out-of-state landfills with virtually no prior notice, and associated piling-up of wastes in the streets ... receiving states have passed executive orders, regulations and statutes targeted at reducing or eliminating the interstate movement of solid waste," the NJDEP wrote in 1993, "it makes little sense environmentally or economically to continue the dependence on out-of-state disposal ... New Jersey's objective is to become self-sufficient in disposal capacity within the next seven years."[30] Flow control measures were the tactic that held the self-sufficiency goal together.

Beginning in 1994, however, this bedrock of waste management planning in New Jersey would begin to crumble. That year, a seemingly minor lawsuit between a recycling collector and a town just north of New York City—*C & A Carbone, Inc. v. Town of Clarkstown*[31]—erupted into a case before the U.S. Supreme Court over the constitutionality of the flow control concept. Like many municipalities in New Jersey, Clarkstown, New York had been forced to close its old landfill on environmental grounds. When it did, it agreed to build a transfer station which would receive all the nonrecyclable waste from the town at a set charge of $81 per ton for at least five years. While the town itself neither owned nor operated the transfer station, town officials planned on purchasing the facility from the private operator at the end of that five-year period, for the price of $1. Essentially, the $81 per ton rate was determined to be adequate compensation to repay the private construction and operation of the facility over the initial five-year period. CA Carbone, Inc., was a recycling processor located in Clarkstown that needed to dispose of the "residues" and nonrecyclable fraction of the material it received at its own facility. CA Carbone was located near the New York-New Jersey border and did, in fact, process some amount of material coming from New Jersey. In March 1991, a traffic accident involving a CA Carbone truck revealed that the company was hauling its residues to a disposal site in Indiana. Subsequent surveillance by the Clarkstown police revealed that this was regular practice, with the company sending waste as far afield as Florida, Illinois, and West Virginia in addition to sites in Indiana. The town of Clarkstown sought an injunction against CA Carbone, claiming that the company was obligated to send residues to the town-designated facility. CA Carbone responded by filing its own lawsuit to enjoin the town's flow control regulations in their entirety.

After months of arguments in district and appellate courts in New York, the case landed before the U.S. Supreme Court. Here, the justices determined that local ordinances like the one passed by Clarkstown in fact violated the dormant commerce clause of the U.S. Constitution. Justices found that Clarkstown's ordinance was an attempt to regulate interstate commerce. In particular, the Supreme Court observed that because Clarkstown (and the state of New York, more generally) had other means by which to pay for the facility, "the ordinance's revenue generating purpose by itself is not a local interest that can justify discrimination against interstate commerce. If special financing is needed to ensure the transfer station's long-term survival, the town may subsidize the facility through general taxes or municipal bonds, but it may not employ discriminatory regulation to give the project an advantage over rival out-of-state businesses."[32]

Justice Anthony Kennedy, in delivering the opinion of the majority, noted that:

> Clarkstown protests that its ordinance does not discriminate, because it does not differentiate solid waste on the basis of its geographic origin. All solid waste, regardless of origin, must be processed at the designated transfer station before it leaves the town . . . the ordinance erects no barrier to the import or export of any solid waste, but requires only that the waste be channeled through the designated facility . . . [But] as the town itself points out, what makes garbage a profitable business is not its own worth but the fact that its possessor must pay to get rid of it. In other words, the article of commerce is not so much the solid waste itself, but rather the service of processing and disposing of it.
>
> With respect to this stream of commerce, the flow control ordinance discriminates, for it allows only the favored operator to process waste that is within the limits of the town . . . The essential vice in laws of this sort is that they bar the import of the processing service. Out-of-state meat inspectors, or shrimp hullers, or milk pasteurizers, are deprived of access to local demand for their services. Put another way, the offending local laws hoard a local resource—be it meat, shrimp, or milk—for the benefit of local businesses that treat it. The flow control ordinance has the same design and effect. It hoards solid waste, and the demand to get rid of it, for the benefit of the preferred processing facility . . . The flow control ordinance at issue here squelches competition in the waste-processing service altogether, leaving no room for investment from outside.[33]

Of course, New Jersey already had considerable experience with the dormant commerce clause, having sued, and having been sued, several times in the 1970s and early 1980s over whether or not state officials could bar the disposal of out-of-state wastes within state boundaries (they could not), and whether the Solid Waste Act empowered state officials, in particular the NJDEP, to create rules directing haulers to use particular disposal sites (it did). Like the transfer station that Clarkstown had built, which used flow control rules as a financing mechanism, flow control was a considerable component of the state and counties' financing plans for waste management facilities. In fact, by 1994, there was already more than $1.65 billion in outstanding bonds, divided among fifty-three separate issues by New Jersey local or county authorities.[34] Disposal fees represented virtually the only revenue stream by which these bonds would be repaid.

Perhaps unsurprisingly, a number of suits were filed specifically targeting New Jersey's flow control rules, all of which gained considerable traction after the *Carbone* decision was handed down. The most significant of these cases was *Atlantic Coast Demolition v. Board of Chosen Freeholders* and the subsequent appeal. Atlantic Coast Demolition was a construction and demolition debris (C&D) processor located in Philadelphia, but with customers in southern New Jersey. Atlantic Coast Demolition filed suit that NJDEP's rules for waste disposal were unconstitutional, for largely the same reasons that CA Carbone, Inc. had claimed in New York. Like Carbone, Atlantic Coast Demolition hauled materials to its own site in Philadelphia, where it would separate out recyclable content and arrange for the disposal of the rest. Also like Carbone, Atlantic Coast Demolition frequently made arrangements to send the unrecyclable part of the waste it collected to landfills much further afield, including several in Ohio. However, one of the rules that the NJDEP had developed over the 1970s and 1980s was that for companies collecting "mixed waste"—recyclable and nonrecyclable wastes together in the same load—while there were no restrictions on the destination of the recyclable fraction, the nonrecyclable component of the waste was required to be disposed of per flow control rules in the county in which it was originally collected. That is to say, Atlantic Coast Demolition was required to drive the nonrecyclable waste back into New Jersey and dispose of it at the designated site, at the predetermined price—or, it could pay the disposal facility a fee equal to the amount it would have paid had it actually used the facility. The company took issue with these rules and filed suit.

The case ultimately unfolded before the Third Circuit of the U.S. Court of Appeals. Per the case documents, neither litigants nor judges hearing the case disputed the fact that New Jersey's waste management scheme was highly complex, that the flow control rules evolved in response to the unique challenges of managing solid wastes in the state, or that flow control was a central component of the state's solid waste management strategy. However, the court articulated that because of the flow control rules, the NJDEP and each county of the state were not only making the rules for waste disposal, but actively participating in the market for waste disposal services and favoring their own facilities. As Justice Walter K. Stapleton articulated in the court's opinion,

> New Jersey participates (or directs local government entities to participate) in the waste disposal market as sellers and purchasers of waste disposal services and disposal capacity. The districts "sell" waste disposal services, according to the Department [NJDEP], through the designated disposal facilities. Where a district has opted not to own or operate the designated facilities directly, it

"purchases" these services for "resale" by contracting with private facilities for the provision of waste disposal services. Thus, the Department maintains, the waste flow regulations simply represent a means by which the state manages the districts' market participation . . .

While we do not quarrel with the Department's characterization of the districts' activities as involving purchases and sales of disposal service and capacity, we cannot agree with its conclusion that the waste flow regulations, therefore, cannot be violative of the dormant Commerce Clause. When a public entity participates in a market, it may sell and buy what it chooses, to or from whom it chooses, on terms of its choice; its market participation does not, however, confer upon it the right to use its regulatory power to control the actions of others in that market . . .

Under New Jersey's solid waste disposal program, the districts are doing more than making choices about what waste they will accept even in those instances where the district owns the designated facility. The waste flow regulations purport to control the market activities of private market participants. Those regulations do not concern only the manner of operation of the government-owned or government-managed designated disposal facilities; they require everyone involved in waste collection and transportation to bring all waste collected in the district to the designated facilities for processing and disposal. They do not merely determine the manner or conditions under which the government will provide a service, they require all participants in the market to purchase the government service—even when a better price can be obtained on the open market. New Jersey's waste flow control regulations were thus promulgated by it in its role as a market regulator, not in its capacity as a market participant. As a result, those regulations are not immune from review under the Commerce Clause.[35]

The state argued that under its rules, out-of-state facilities could in fact be designated the "official" disposal site of a county, so long as the county went through the correct planning process. But in the opinion of the court, this fact "does not transform a fundamentally discriminatory scheme into a non-discriminatory one . . . the Department acknowledges that it approves district plans only if they are consistent with the 'core' goal of having all of New Jersey's solid waste processed and disposed of in New Jersey . . . This can be accomplished, and is being accomplished, only by selecting existing and proposed instate facilities whenever possible."[36]

Even as the *Atlantic Coast* suit played out, additional suits were filed against New Jersey and the NJDEP. In one of them, *Waste Management of Pennsylvania, Inc. v. Shinn*, a U.S. District Court in New Jersey concluded that the state's goal of self-sufficiency in waste disposal in and of itself discriminated against interstate commerce. In the course of that case, details about denials of permits emerged, showing that NJDEP officials were generally barring counties from making any sort of arrangements for out-of-state disposal of their wastes, and pointing specifically to the agency's "self-sufficiency doctrine" as the rationale.[37] District Judge Joseph H. Rodriguez noted in the court's opinion, that "New Jersey's self-sufficiency policy is facially discriminatory against out-of-state waste disposal competitors" and that "New Jersey's Statewide Plan expressly provides that new proposals or contracts for the long-term use of out-of-state disposal facilities by New Jersey's counties will not be approved." He continued that "In furtherance of New Jersey's stated self-sufficiency mandate, all long-term waste disposal contracts which provide, even in part, for out-of-state waste disposal were rejected to the extent that such out-of-state disposal continued beyond December 31, 1999" and then offered examples of such rejected proposals involving Atlantic, Cape May, Mercer, Morris, Hudson, and Essex Counties.[38]

Furthermore, the state's own position on market development for recycled materials played a role in the district court's decision, since flow control rules very conspicuously did not apply to recyclables:

> The State Defendants cursorily assert that self-sufficiency serves such goals as "the assurance of long-term disposal capacity, the encouragement of source reduction and recycling, the proper closure of sanitary landfills, and the composition of waste received for incineration." However, the DEP's asserted interests are unsupported by the record and are insufficient to justify an outright ban on export of New Jersey's solid waste after December 31, 1999 . . . The State Defendants have failed to demonstrate that long-term disposal capacity can *only* exist within New Jersey. Furthermore, source reduction and recycling are not dependent upon whether the disposal facility is located within New Jersey's borders. In fact, as to recyclables, the DEP has stated that "the interstate movement of recyclable commodities must remain unaltered by legislative or regulatory restrictions to maintain the free market system of commerce and to maximize opportunities for the marketing of materials."[39]

Ultimately, the court found no part of the state's argument satisfying, and struck down flow control as a tactic for implementing the self-sufficiency doctrine:

> The DEP has stated that "the ultimate point of self-sufficiency [is] responsible disposal of New Jersey's solid waste with as little impact as possible on other communities." As a factual matter, there is no evidence in the record to support the proposition that New Jersey could not achieve this goal by less restrictive means. We find unpersuasive each interest that the DEP has offered to justify a self-sufficiency policy which prohibits the free flow of New Jersey's solid waste to out-of-state facilities . . .

The self-sufficiency mandate of the Statewide Plan creates a barrier to solid waste exports and the use of out-of-state disposal facilities . . . The self-sufficiency mandate discriminates against out-of-state waste disposal facilities in favor of in-state economic interests without serving any legitimate local interest which could not be addressed by less restrictive means . . . New Jersey's self-sufficiency policy favors in-state disposal providers and ultimately blocks the flow of solid waste at its borders . . . Accordingly, this court holds that New Jersey's policy of waste disposal self-sufficiency violates the Commerce Clause.[40]

Reactions to the rulings were swift and varied. Public officials involved in devising the state's waste management system, generally speaking, interpreted the rulings as a devastating blow to the infrastructure that had been built up over the past quarter century. Gary Sondermeyer, former chief of staff at the NJDEP and closely involved in drafting many of the state's waste management policies during the 1980s, observed that the *Carbone* and *Atlantic Coast* decisions "brought down the whole waste system in New Jersey."[41] In contrast, John Turner, a vice president at Browning-Ferris Industries (BFI, a major national waste hauling firm), hailed the courts' decisions as supporting the liberties enshrined in the U.S. Constitution. He wrote in the *Villanova Journal of Environmental Law* that the *Carbone* decision in particular was "simply the latest in a series of decisions that have questioned measures that seek to 'hoard' waste volumes or to exclude non-local wastes. Through consistent application of the principles that underlie the Commerce Clause, courts can ensure that the goals of the founding fathers are promoted and that a national market for waste services is fostered."[42] Others held a more measured, but still positive assessment of the courts' rulings. State senator Henry P. McNamara shared in one of the first hearings held in the post–*Atlantic Coast* era that "I believe there is a great opportunity for New Jersey if we act decisively and wisely . . . Maybe the Supreme Court did us a favor when it decided the *Carbone* case, perhaps even when the Third Circuit decided the *Atlantic Coast* . . . It gave us the opportunity to craft a new system that is less costly, more efficient, more fair, and more environmental."[43]

To be sure, flow control had long had its opponents within New Jersey. Critics charged that flow control artificially raised the price of disposal in the state—which it most certainly did—meaning that in the long term it cost New Jersey residents and taxpayers much more to deal with their waste than it would have otherwise. In 1993, at the height of state planning and operation of flow control rules, the average disposal fee ("tipping fee") at New Jersey's landfills, WTE facilities, and transfer stations was $93.[44] In 1997, after the *Atlantic Coast* decision, the average tipping fee in New Jersey had plummeted to around $55— with no magic increases to the supply of disposal capacity available in the state to explain the change in price.[45] Since flow control was intended to recover the costs of constructing, operating, and financing facilities, disposal fees were set to achieve those outcomes rather than to be competitive with other disposal sites. Furthermore, because counties had pursued different strategies (with different costs) for meeting their disposal obligations, disposal rates frequently varied from one county to the next. There was tremendous variation in tipping fees across the state. In 1993, for instance, it cost $116 per ton to dispose waste in Sussex County, but just $49 per ton in Burlington County.[46] Even as various attempts to *equalize* rates across the state were explored, many remained unconvinced that flow control could ever result in prices that were perceived as fair, let alone competitive with out-of-state disposal options. Thus there were enticements to skirt flow control rules both across county lines, and across the New Jersey state line as well.

While the rules for flow control were clear, enforcement efforts were typically weak and inconsistent. Albert Fiore, then the executive director of the Hudson County Improvement Authority, testified during a 1989 hearing that:

> The system is failing because of migrating waste flowing freely to cheaper destinations. The solution is not more study, nor achieving greater predictive powers. It is simply to appropriate sufficient enforcement monies to do the job . . . The penalties are laughed off as petty cash expenditures by the violators. Put harsh teeth, including confiscation, into state and local enforcement and the need for more studying will dissipate . . .
>
> Rate averaging will not achieve the desired purposes. It's a fallacious premise because New Jersey is not a separate planet. Even after you have fashioned an "equitable" averaged rate, migratory waste will continue to flow into and out of the state. It will always come in from New York, where the costs are high, and flow to Pennsylvania and points west and south, where the costs are lower. Unless, you employ those aforementioned enforcement tools, this will always be the case.[47]

Under flow control rules, New Jersey haulers could not legally utilize lower-cost alternatives in other states or even in neighboring counties. Flow control rules created a captive audience, a fact which many in the waste industry resented. It is not surprising, then, that many haulers did their utmost to skirt flow control rules or flouted them entirely, and absorbed the fines and penalties for flow control violations as simply a cost of doing business, as Fiore noted. By 1988, state officials estimated that "up to 20 percent of the waste that should have been going to country transfer stations was being shipped directly out of state in contravention of waste flow orders," with some counties—Essex, Passaic, and Bergen in particular—receiving 51 percent, 74 percent, and just 39 percent of their expected waste volumes, respectively.[48] A State Commission of Investigation report relayed the story of one waste hauler who had more or less never adhered to flow control rules: "Cheating on waste flow orders gives the greatest competitive advantage to the collectors that are the most flagrant violators; and violations are, indeed, brazen and widespread. For example . . . the [Board of Public Utilities] revoked the license of Fiorillo Brothers of New Jersey, Inc., and barred five principals of the firm from the solid waste industry in New Jersey based, in part, on violation of waste flow orders. The company, which serves commercial customers throughout northern and central New Jersey, had, according to the BPU, continued to serve all of its accounts; yet . . . it had not shown up at a single authorized transfer station."[49]

Alternatively, state-level investigations noted how lack of coordination between administrative departments often put waste management firms into untenable situations, where violating the flow control orders was the only way to stay in business. For instance, a 1989 Commission of Investigation study of violations of New Jersey's waste management regulations described the frequent situation where "abrupt changes in waste flow orders, usually when a landfill is closed or its volume severely limited, have resulted in hardship for individual haulers . . . [and] 'rate shock' for their customers." To elaborate, "If, for example, a change in waste flow requires a collector's trucks to travel greater distances or to wait in lengthy lines at a different disposal facility, the collector may have to operate more trucks in order to adequately service its existing customers. However, the . . . process for approval of additional trucks may take up to several weeks and the collector cannot pass increased costs on to its customers in the meantime."[50]

Some of the harshest criticism of flow control came from New Jersey politicians who had firsthand experience with flow control rules, having held office at municipal or county levels during the years of flow control, and then secured higher offices during the period of post–*Atlantic Coast* reform. One of the most vocal critics was state Assemblyman John E. Rooney. Rooney was formerly mayor of a small town in Bergen County and also served on the Bergen County Utilities Authority. He related how, in his experience, seemingly the entire apparatus of state government conspired to artificially inflate waste management costs and generate the huge mountain of debt facing counties:

> what happened when the solid waste issue came up in [the early 1980s], the State had said we had to find an in-state solution . . . We had to come up with—each county had to come up with a solid waste solution. At that time, I was a Commissioner on the Bergen County Utilities Authority, and any solution that was submitted to the state that was not an incinerator or . . . did not involve sharing of an incinerator was rejected by the State. You had to have somehow an incinerator in your plan. So if they are saying that they didn't mandate it, they are wrong . . .

But that's what happened, those are the facts. We all went down this path, and we went down it like the lemmings. We all did it at once. At one point in time we all marched over that cliff. And the problem was that when you set up a scenario that says you are closing all the landfills in the State, you all have to go out for incinerators at the same time. What happens? It becomes a seller's market. Whether it was solid waste landfill space, whether it was incinerators, we paid top dollar. We paid more than top dollar, we got ripped off. And that's what happened, and it was ... mandated by the State ...

Further, every plan, every plan amendment that went through, yes, went through your utilities authority, yes, went through your county, but had to be approved by the State. If it was wrong, if it was something that was excessive, the State should have picked it up. Every budget that every utility authority had went through the State Division of Local Government Services. Every bond issue that ever was passed went through the State Department of Local Government Services.

The State has had its big fat fingerprints, footprints, every kind of print imaginable on every step of solid waste mismanagement in the state since the early '80s. That's what this is all about.[51]

In sum, it was clear to anyone involved in New Jersey's waste management system, whether as market participant or regulator, that for all of its positive and innovative qualities, the scheme which had evolved since the 1970s was far from perfect. Discussions over reforming flow control in New Jersey unfolded over the final years of the twentieth century, through the course of dozens of hearings, proposals, and legislative maneuvers.[52] The rulings against New Jersey's flow control system gave the state a short period of time to rewrite its regulations in a manner that increased competition for disposal services. The process consumed all of the oxygen relating to waste management in New Jersey during this time. So many different proposals were heard that it is preferable to focus on the actual outcomes and then retrace some of the steps taken to arrive at the destination.

The upshot of years of legal wrangling was a system where towns, counties, and sometimes counties acting as agents for a group of towns, had to procure waste management services in the same fashion as they would other types of contracted services. The ultimate menu of options available to counties in the wake of *Atlantic Coast* thus consisted of the following, per the NJDEP:

> *Non-discriminatory Bidding Flow Control*: Under this system, as a result of a non-discriminatory bidding process, which allows in-state and out-of-state companies to bid on a contract for disposal of a county's waste, counties can institute solid waste flow control on the waste contracted. The waste that is subject of the contract is required to be disposed of at the contracted location under penalty of law.

Intrastate Flow Control: An intrastate flow control system mandates that all non-recycled solid waste generated within a county which is not transported out-of-state for disposal shall be disposed of at the designated in-county disposal facility.

Market Participant: A market participant system allows a county-owned facility to compete with other in-state and out-of-state disposal facilities for the disposal of the solid waste.

Free Market: A free market system allows solid waste generated within a county to be disposed at whatever disposal facility agrees to accept the waste, based on terms freely agreed to by the generator, the transporter and the disposal facility operator.[53]

Ironically, the outcome of the forced reform was not always so radically different from what had existed prior to the *Atlantic Coast* decision. As the 2006 *Statewide Solid Waste Management Plan* pointed out: "It should be noted that since 'Atlantic Coast' and the end of *state* regulatory flow control, a number of *counties* have undertaken constitutional re-procurement of their disposal needs in a manner that allows them to control the flow of waste and therefore their management of it. In addition, there are several counties that have instituted intra-state flow control plans. Those plans allow for the free movement of waste out-of-state; however, if the waste stays in state, it is directed to a facility in that county."[54] Some counties, like Ocean, Morris, and Mercer, had in fact designed waste management plans that never fell afoul of the *Atlantic Coast* decision in the first place. These counties—at times to the great irritation of officials at the NJDEP, in the case of Morris county[55]—had at the core of their plans a fairly competitive bidding and contracting process to begin with, in some cases, ironically, as a result of refusing to engage fully with the state's planning procedures. Ocean County had long contracted with a privately owned landfill, and Morris County with transfer station operators or even landfills themselves (some out of state). Mercer County, while attempting to plan various disposal facilities within its borders, had been contracting directly with various Pennsylvania landfills just across the Delaware river. While some aspects of these counties' systems were rebid in compliance with the *Atlantic Coast* decisions, complete reformation of the counties' waste management plans was largely avoided since, for the most part, county government was not acting to both regulate and participate in the market for waste disposal in the first place. It also turns out by taking a different tack in their waste management planning, that Ocean and Morris accumulated considerably less debt in the development of waste management systems than many of New Jersey's other counties and the Meadowlands District.

By 2006 much of the dust surrounding reconfiguration of county waste management plans had settled. Five counties (Gloucester, Hudson, Mercer, Morris, and Union) had developed "non-discriminatory bidding processes" where it could be determined that the waste management disposal contract for the county had been designed and selected in such a way as to have been open to facilities located outside the county. Three counties (Cape May County, Monmouth, and Ocean) adopted "intrastate flow control" rules, which dictated that any waste generated inside each county that was *not* headed to a disposal facility outside New Jersey had to go to a designated facility within the county. Eight counties (Atlantic, Burlington, Cumberland, Hunterdon, Middlesex, Salem, Sussex, and Warren) adopted "market participant" strategies, wherein the publicly owned disposal facilities in the county competed with disposal facilities elsewhere for the business of disposing wastes. Three counties (Bergen, Passaic, and Somerset) opted out of flow control rules entirely, electing a "free market" system where individual municipalities within the county all individually contracted for their waste disposal needs with any facility, located anywhere. Finally, Camden and Essex counties developed hybrid "market participant" and "non-discriminatory bidding" strategies, depending on the types of material in question.[56]

Reconfiguring the specifics of county waste management plans and flow control rules was arguably the easy part of the court-imposed reforms. The question of how to handle the more than $1.6 billion in public debt that had been generated in the process of planning, siting, and building New Jersey's landfills, WTE facilities, and transfer stations during the 1980s and 1990s was far more pernicious. The specter of "stranded debt"—debts with no apparent revenue stream by which to repay them—and even default threatened the credit ratings of various New Jersey counties, utility authorities, and seemingly the state itself. After the initial *Atlantic Coast* decision, the Moody's Investors Service credit rating firm downgraded five counties' rating to "BA"—below investment grade. As Christopher Mushell, the Moody's representative who appeared to testify before the state Senate Environment Committee in 1996, noted, that "Unless alternative mechanisms are established to replace the loss of legal flow control, systems may be in jeopardy of losing an already narrow revenue stream to repay debt and, consequently, may result in further credit deterioration." Mushell warned lawmakers that "Moody's is concerned over how these systems will remain competitive, economically feasible, and politically viable, and generate sufficient revenue to satisfy debt requirements."[57]

In 1999, representatives of nearly every New Jersey county appeared before a state Assembly Solid and Hazardous Waste Committee to discuss the impacts of the *Atlantic Coast* decision on their operations and their finances. A closer look at the experience of Camden County is illustrative of the complexity that many counties faced. Appearing before the committee was Frank Giordano, executive director of the of the Pollution Control Financing Authority of Camden County (PCFA). The PCFA was an entity spun out of the Camden

County government specifically responsible for the financial obligations of the county's solid waste management infrastructure.[58] Camden County had recently built a state-of-the-art WTE facility which commenced operations in 1992, receiving between 400,000 and 500,000 tons of waste each year. As Giordano explained to the lawmakers present,

> Prior to the US Supreme Court's denial of New Jersey's appeal [of the *Atlantic Coast* and related rulings], the PCFA system was charging $94.01 a ton for all waste . . . During 1998 our tipping fees averaged about $47.50 a ton, or a little more than half of the previous charge. When you review the revenues of the PCFA . . . For the six years proceeding 1998 our revenue stream averaged about $56 million a year. Last year we collected $32.5 million . . . we continue to operate with less revenue than we need to meet our financial commitments. And to make matters worse, our debt repayment schedule spiked this year. Last year we had a debt obligation of about $16 million, and this year it increases to a whopping $27 million . . .
>
> What does this all mean in a practical sense? Our debt total is about $185 million, including the approximately $21 million of State debt. We're not collecting enough money to pay both our operating costs and our debt obligations. We have debt service payments due on June 1, 1999 and December 1, 1999. Our debt reserve should be sufficient to meet our June payment but most likely will not be sufficient to meet our December payment. At which time, unless there is assistance, the PCFA of Camden County will default . . . While Camden County has two very desirable solid waste facilities that are convenient and efficient, there is too large a debt to be able to pay for them in a competitive environment. Our remedial investigation process has lead us to believe that without State assistance there is no workable solution.
>
> While this problem is very complex, the basic elements are clear and simple. The number of stakeholders make remedial movement difficult. Re-procurement of waste disposal services [through "non-discriminatory bidding" or "market participant" approaches] does not, by itself, provide for continuing solid waste disposal and repayment of existing debt. The county does not feel that it's in its best interest to assume a repayment schedule responsibility. That leaves the leadership role to the State of New Jersey. Without State direction and assistance, it seems that the question is not if the PCFA of Camden County will fail, but when.[59]

Assemblyman Rooney responded in dismay:

> Twenty-six million dollars at the end of the year. If they don't have it, then they are going to have to default on it. And that's what we are looking at in not only

Camden, but several other counties are into this. Some of them are luckier than others where they can actually go for a while and use some other alternatives . . . I mean you've done a great job to try and survive up to this point.

What we've got to do is try and help them . . . We're basically saying to the [local government], "Screw you, go raise your property taxes." So this is what's happening. The buck has been passed all the way down to the bottom, and we're down on the bottom saying, no, you're not going to do it, we're passing it back up to you. The State of New Jersey has got to step in. This is a critical juncture that we have in solid waste.[60]

According to a study conducted by the NJDEP years after the *Atlantic Coast* decisions, "Counties . . . that expended public funds to construct facilities could not . . . easily modify their systems and still pay the debt incurred. Their rates were generally higher than many out-of- state facilities, due to factors such as availability of open space and density of population, the inability to reject unprofitable portions of the waste stream, and various taxes and surcharges designed to pay for recycling programs and ensure the proper closure of land-fills."[61] That is to say, before *Atlantic Coast* a considerable proportion of many counties' tipping fees was dedicated to costs unrelated to the operation of the disposal facility, but rather servicing debt, supporting recycling programs, and generating a flow of funds escrowed for future pollution prevention efforts. To illustrate the dire impacts on some counties' finances of removing flow control, New Jersey Department of Community Affairs commissioner Jane M. Kenny testified during a hearing immediately following the *Atlantic Coast* decisions that,

> charges for disposal of municipal solid waste currently range from a low of $49 per ton in Burlington County to a high of $125 per ton in Hunterdon County. Debt service coverage is obviously a part of that charge, and the debt service component of the tip fee ranges from $1.10 per ton in Morris County to $43.53 per ton in Sussex County, depending upon the type of system the solid waste district has in place.
>
> Forty-six percent of the debt outstanding is supported either by county deficiency agreements or insurance. Fifty-six percent of the debt would have to rely solely on tip fee revenue. In an environment where flow control does not exist, the State must be concerned about these districts receiving sufficient revenue to cover the debt.[62]

To overcome the revenue shortfall, many counties sought to impose an environmental investment charge (EIC) on every ton of waste *generated* in their county, regardless of where it was disposed. The EIC would be intended specifically to help counties pay off remaining waste management debt, and most were proposed to sunset after some period of time when the debts were paid. Since New Jersey had for many years been applying per-ton taxes anyway, for instance to fund the state's recycling programs, the concept itself was not new. But in the new era of market competition for disposal services, many feared that EICs would once again create distortions. For instance, as Assemblyman Rooney described in his experience with his hometown in Bergen County:

> I can tell you from my own experience and my own county, as both a mayor and a legislator, there is no way that they [the Bergen County Utility Authority] are going to collect the $26 a ton fee that they intend to charge. They are calling it a user charge; however, as a nonuser in my own community—We have gone out for contract for disposal, and ... my disposal right now goes to, of all places, Clarkstown, New York ... We are at a very competitive rate. We went out to bid twice, and then we negotiated the contract after that. This year we are paying $54 a ton flat. The Bergen County Utilities Authority is offering $54 a ton or roughly around there. But if I went there, I'd have to pay another $26 a ton. They are saying, even if I don't go there I have to pay $26 a ton. Well, wrong.[63]

Furthermore, many felt that the EIC approach was politically impossible and even morally questionable. Assemblyman Rooney argued forcefully against the concept, noting: "Let's not kid ourselves. We are talking about a tax that is going to be put on garbage, or whatever, for all of our property taxpayers to absorb. It's wrong ... You are going to see either a county tax increase or a municipal tax increase if EICs are passed, and that's not acceptable."[64] Rooney, and others, warned that disposal rates could actually end up higher under a system with EICs than they ever were during the height of flow control:

> In fact, I will give you one example why the EIC is so wrong. The EIC that they are talking about in my county and I think a couple of the other counties are saying that the municipality has to collect an EIC for the tonnage generated by the commercial and industrial generators. We never had anything to do with the commercial-industrial generators of garbage in our towns. Now, in my town, for example, I have 2000 tons of residential waste. They [the NJDEP] tell me I have 3000 tons of commercial-industrial waste. I would have to pay $26 a ton [EIC] on not 2000 tons, but 5000 tons of waste. I wind up paying $119 a ton when the worse-case scenario, the last time we had flow control, was $102. We'd actually pay more under this new scenario.[65]

Counties that had avoided accruing large amounts of debt to build disposal facilities or transfer stations—Ocean and Morris in particular—were opposed to the EIC concept as well as any form of restructured financing of waste debt

that could require all twenty-one counties to shoulder the burden. As the director of the Ocean County Department of Solid Waste Management, Alan
Avery, testified to one assembly committee, "Our situation in Ocean County
has not changed much since flow control was declared unconstitutional . . . debt
was not incurred in constructing our new facilities. This continues be the case
today. Since we are debt free, there has not been any consideration of implementing an environmental investment charge . . . since Ocean County has carefully managed the cost of its solid waste management program, it is reluctant
to assume the burden of financing other county solid waste management
decisions."[66]

Various iterations of the EIC concept were batted around for years, but were
never implemented at the scale as to fully recover the revenue lost from the
drop-in tipping fees precipitated by *Atlantic Coast*. Ultimately, the state of New
Jersey did step in to assist county governments meet their debt obligations,
albeit in an ad hoc, piecemeal fashion. In the four years inclusive of 1999 and
2001, state lawmakers appropriated $80 million to subsidize counties' debt pay-
ments relating to solid waste, "intended to provide short-term financial assistance to select counties and authorities and that had difficulty making debt
payments through the collection of their respective tip fees."[67] Also, in 1998,
New Jersey voters approved a "public question" allowing state government to
simply forgive over $100 million in public solid waste loans. Several other leg-
islative solutions were considered, including proposals to "equalize" debt
payments across all county governments—including those who had no solid
waste-related debts—and to simply have the state pick up the tab for all outstanding waste-related debt. None of these proposals were adopted in the years
following the *Atlantic Coast* rulings.

Despite the limited actions that state government had taken, in 2003 there
was still over $930 million of county solid waste debt outstanding, with five
counties—Bergen, Hunterdon, Morris, Ocean, and Somerset—having eliminated their waste debts entirely. Mercer County had the most outstanding
waste debt, at nearly $140 million, and Salem County the least at about $6 million.[68] By that same year, over $107 million of debt had been forgiven through
state programs and millions more reduced by direct subsidy from the state.
Essex County had seen the most debt, some $44 million, wiped off its books,
even while several counties, including Mercer and others with debt loads large
and small had received no forgiveness.[69] During this first part of the twenty-
first century, New Jersey's counties have been continually working to pay down
remaining debts, with on-and-off assistance from the state of New Jersey, fluc-
tuating with both changes in government and turmoil in the state's finances.
In some instances, counties have sold or leased the assets on which they are paying
debt as one strategy for reducing the debt burden. In other instances, as in
the case of Camden County in 2010, the state permitted the county or waste

management authority to tap into emergency reserve funds previously set aside for other purposes in order to avoid default. In nearly every case, however, counties having considerable waste management–related debts have faced a long and difficult process of recovering from the losses of revenue imposed by the *Atlantic Coast* decisions. Every dollar that went toward debt service, especially in light of reduced revenues, was one less dollar that counties could invest in improvements and even actual operations of solid waste facilities. It is no surprise, then, that the end of the flow control system marked the beginning of a period of stagnation for New Jersey's solid waste infrastructure.

New Jersey's Wastes in the New Millennium

The end of flow control, and subsequent dismantling of the state's approach to facility planning, financing, and operations that had been built up over the preceding thirty years, sapped much of the momentum and innovation from waste management in New Jersey. This is perhaps most clearly reflected in figures for recycling. In 1990, for instance, New Jersey recycled around 34 percent of its municipal solid waste.[70] The figure peaked in 1995, when the state reported a municipal solid waste recycling rate of 45 percent. In 1996, in addition to discerning the impacts of *Atlantic Coast* and formulating a response to the courts' rulings, a fiscally conservative state legislature allowed the "landfill tax" to expire as part of the reforms. As Michael Hogan, an attorney for the NJDEP, testified at that time, "Currently, our tipping fees do more than just pay for the disposal of solid waste. They provide for recycling. They provide for hazardous waste management. They provide for research and development in things such as methane control and landfill mining and educational opportunities. These are things that New Jersey, as a small, but the most densely populated State, needs to continue to pursue."[71]

Accordingly, by 2003, the municipal solid waste recycling rate had declined to 32.7 percent, as "the loss of the program's dedicated state funding source [the landfill tax] as well as the declining solid waste disposal fees that resulted from [*Atlantic Coast*] . . . played major roles in this decline" according to an NJDEP analysis.[72] The statewide municipal solid waste recycling rate has fluctuated between 33 percent and 44 percent ever since, and while impressive, remains some distance from the 50 percent target that had been established by state planners.[73] Even as the landfill tax was reinstated in 2008 via the Recycling Enhancement Act, this situation has worsened considerably in the years since the Chinese government has halted imports of recyclables from the United States, a matter discussed further in chapter 6. Little insistence on self-sufficiency in waste disposal and processing remains.

All this is to make the case that in the years since flow control responses were settled, little of importance has happened with New Jersey's waste management system. To be sure, some of that is because of the massive stranded debts faced by some counties, who were, and remain, hesitant to pursue new facilities and programs of any type. More fundamentally, the courts' striking down the state-directed flow control concept has removed the possibility of a strong, centralized manager in waste planning. During the heights of flow control, New Jersey officials developed a centralized plan and took steps toward implementing a more or less coherent vision for waste management in the state. While this approach was problematic in many ways, not least the economics of disposal and the nightmares of public financing that it introduced, it nevertheless attempted to operationalize a singular and consistent vision. For instance, even though it was specifically stricken by the courts, the notion that an entire state would take responsibility for all the wastes it produced via an official policy goal of self-sufficiency is inspirational, even heroic, and also environmentally progressive—and unlikely to ever be seriously considered again now that wastes are allowed to freely flow to wherever the lowest disposal price can be found. In the words of Gary Sondermeyer—formerly chief of staff at the NJDEP and deeply involved in many of the state's waste management planning efforts during the 1980s and 1990s, all of New Jersey's responses to waste management prior to *Atlantic Coast* had been "extraordinarily, extraordinarily bold."[74] Those taken since have been, somewhat out of necessity, less so.

Yet, in reflecting on the history of New Jersey's waste management system, Sondermeyer observed,

> taking a step back, my hindsight is that it was successful. We took a totally arcane way of dealing with waste that absolutely cost hundreds of millions of dollars if not billions of dollars to clean up and we did put in place probably, arguably, the most comprehensive waste system in the country. Because we have had mandatory recycling for 30 years, we are one of the highest recycling states in the country. Our infrastructure which we built through that bizarre wild, wild west through being upheld in the courts to being struck down in the courts, to being re-imposed by the courts . . . We built 12 modern well-run landfills. The incinerators are phenomenal, phenomenal . . . extremely well-run, so I mean you can make an argument that all of this pain and all of this angst and all the trials and tribulations worked.
>
> To me, the really interesting thing is what will happen next . . . landfills keep getting expanded. So we haven't built any new thing . . . since the late 1980s . . . At some point in time, at some point in time, these landfills will be full. At some point in time, these mass burn incinerators [will be retired]. At some point—the question will become, "What will happen?"[75]

In the early years of the twenty-first century, waste management planning in New Jersey has been less coordinated than in the past, with counties pursuing their own operational, financing, and environmental strategies amid occasional bursts of activity from the state. For instance, various committees and even legislation to address food waste, electronic wastes, composting, and other dimensions of the waste stream have been explored by different components of state government, but with very few actionable results.[76] Most counties have attempted to implement one response or another to these new waste management challenges, but the majority of these have been limited-run pilot programs with uncertain futures. This is not the fault of any particular actors—state, county, municipal, or private—but perhaps due to the lack of a unifying vision for waste management and limited mechanisms for creating a coordinated, statewide response and funding new infrastructure. County governments in New Jersey have neither the financial firepower nor willingness to cooperate with one another to create meaningful, comprehensive solutions to waste management challenges. Though it was not a flawless performance, the state of New Jersey once played this central role. The best example of coordinated, state-led action in the waste management sphere was perhaps the emergency response to 2012's Superstorm Sandy when, in a matter of days, immense volumes of wastes were generated by a natural disaster. However, many of the actions taken at that time were intended simply to recover from the storm's impacts, and included statewide measures issued under emergency conditions.[77]

Given the diminished role the state of New Jersey now plays in setting a waste management agenda, it is clear that lasting improvements to what has become a stagnant system will have to come from another direction. Market-based reforms are the most likely to pass the tests of the Commerce Clause and attract participants, and are the subject of the final chapter.

6

Conclusions and
Looking Forward

● ●

> The solid waste industry has become
> increasingly complex and no one report
> could ever pretend to be exhaustive of
> the regulatory problems the industry
> presents. However, we believe that the
> foregoing presents a very solid frame-
> work upon which the legislature can
> draft new legislation and regulations
> which will more directly confront the
> issues raised herein, and ensure that New
> Jersey consumers pay a fair price for solid
> waste collection and disposal in the
> future.
> —1981 Report of the Task Force to Study
> Solid Waste Regulation

In early August 2019, FCR Recycling LLC sent a letter to its client municipalities in the northern part of Camden County, informing them that the company would be imposing strict new penalties for receiving loads of recycling that
did not meet certain standards.[1] "We are at a point where we must enforce the
contamination (dirty) standard rigorously," the company wrote, before explain-
ing that if the company rejected any given load of materials, "the offending

municipality has two hours to pick up the contaminated load, or agree to have FCR either take the load to a solid waste facility, i.e. the landfill (paying $250 in handling charges in addition to the actual trash cost plus a 15 percent surcharge to transport) or have FCR process the load (if not too contaminated) and pay an additional $75 per ton."[2] The notice prompted surprise and consternation among officials of the client towns, who debated the merits of suing FCR because the penalty provisions were not included in the original contract. Nevertheless, few could dispute the underlying motive for the penalty protocol: the United States' market for most recyclable materials had collapsed. The impacts of China's "green fence" and "national sword" policies implemented a few years prior restricting, and eventually banning, imports of foreign and particularly U.S. "postconsumer materials" (recyclables) had by 2019 finally percolated down to towns and counties in New Jersey. Recycling processors had lost their single biggest market and remaining buyers could be more selective about the quality and types of material they would purchase, restrictions ultimately changing the way intermediate firms like FCR would do business. In some places, like the service area of FCR in northern Camden County, new penalty protocols were introduced. In other areas, a substantial price differential emerged between trash disposal and recycling processing. For example, around the same time as FCR's announcement, the town of Colts Neck in Monmouth County reported prices of $86 per ton to dispose of garbage and $135 per ton to manage recyclables.[3] Stringent penalties and inflated spot prices, however, were different reactions to the same problem. New Jersey's once lauded recycling system was broken.

Yet nearly twenty years into the new millennium, the collapse of recycling markets was not the only issue confronting waste management in New Jersey. Some disposal facilities around the state were starting to show their age and dysfunction. For example, the waste-to-energy (WTE) facility in Camden, in a sign of decreasing efficiency, was rated the second-largest emitter of lead of all such facilities nationwide.[4] Other facilities proved, somehow, to be uneconomical despite still-rising volumes of New Jersey wastes needing disposal. Although being "run consistently well," the Warren County WTE facility was slated to be mothballed in mid-2018, due to "challenging local market conditions."[5] The vast "Edgeboro"/Middlesex County Landfill complex in East Brunswick—reaching its originally designed capacity in 1987—received in 2016 an approval by local officials for an additional ten years of operation through 2030. Per one local news report from 2019,

> Around the same time that the privately-owned "Edgeboro" landfill reached its capacity in 1987, the Middlesex County Board of Chosen Freeholders voted to place the adjacent county-owned landfill under the purview of the MCUA [Middlesex County Utilities Authority]. Soon after, the MCUA took over the

private dump through eminent domain, securing the right "to landfill solid waste around, next to, and over the top" of the old landfill, according to state documents. Presently, the expanded landfill site collects four million pounds of trash every day from communities around the county . . .

. . . the Middlesex County government amended its solid waste management plan in 2016 to allow for dumping in the landfill to continue past the year 2020, opting to extend its life until 2030. Officials have no public plan for what will happen after that. The move will ultimately allow the county to cram an additional 7.9 million cubic yards of waste into the existing footprint, causing the massive landfill to rise some 55 additional feet to a height of 220 feet above sea level.[6]

Around this same time, nearby residents had registered nearly 200 complaints about a "rotten egg" odor emanating from the site in just a three-month span between November 2018 and January 2019. Facility officials insisted that the smell was, ironically, associated with construction and installation of a new odor control system and that the unpleasant scents were not linked to any health hazards. But nearby residents took little solace in such assurances, instead raising concerns about the future impacts of piling still more trash into a landfill already appearing to burst at its seams.

Features of New Jersey's waste history that many considered dead and buried were also starting to reappear. Elements of organized crime once more asserted themselves, no longer (directly) in the waste hauling business but in more obscure corners of the industry. One of the biggest scandals was in the world of "dirty dirt," contaminated soil from industrial sites being resold as clean fill for construction projects or even as soil amendments for farms. In 2017, a "dirt broker" with ties to organized crime entities in New York City "was caught passing off waste contaminated with toxic chemicals as clean fill and selling it to be used for a residential and commercial development in Old Bridge."[7] This was just one of many similar schemes documented by a State Commission of Investigation study of corruption in the construction and demolition debris sectors.[8] While the "A-901" laws had attempted to root out organized crime from waste collection activities in New Jersey, the commission in two reports from 2016 and 2019 "revealed significant loopholes in the oversight and regulation of Class B recycling that have enabled unscrupulous 'dirt brokers' and others to pose as legitimate recyclers and thus escape licensing requirements and basic background checks like those required for individuals engaged in solid waste operations."[9]

Perhaps nothing so neatly encapsulates the problems facing waste management in New Jersey twenty years into the new millennium as the liquid waste disposal scandal at the Keegan Landfill in Kearny, in the Meadowlands area. Despite having been slated for closure numerous times since the 1960s, in 2019

Keegan Landfill remained the only active disposal site in the Meadowlands area. The 110-acre site, like Edgeboro, received continual approvals to extend its working life. In fact, one of these approvals came from the New Jersey Superior Court, which ruled in 2016 that closing the site would have "drastic and deleterious effects on surrounding communities and taxpayers."[10] It is undoubtedly true that Keegan Landfill had become part of the economic fabric of the Meadowlands, collecting in 2018 nearly $18 million in disposal fees. However, given its status as the only remaining landfill serving the close-in suburbs of New York City, certain practices had reverted to the "bad old days" of the early twentieth century when anything and everything was dumped at the site. In August 2018 the facility received a violation notice from the New Jersey Department of Environmental Protection (NJDEP) for illegally accepting—during an inspection visit from regulators—"liquid sewage sludge material" from a nearby wastewater treatment plant. New Jersey landfills had been prohibited from accepting sewage sludge since 1985, but Kearny mayor Alberto Santos alleged that disposal of sewage sludge was typical operating procedure at Keegan Landfill, and demanded a criminal investigation.[11] "It was a very orchestrated operation," said Mayor Santos; "they were accepting an illegal substance, and there is reason to believe they have been doing it for quite a while."[12] Like residents near Edgeboro, people living near Keegan Landfill had recently been complaining about severe sulfurous odors, which Mayor Santos posited as connected to the disposal of sewage sludge at the facility. Being the age of social media, the alleged dumping was captured with a cell phone video and posted to Facebook.

In April 2019 Keegan Landfill operators entered into a consent agreement with the NJDEP to supply more frequent environmental testing results to the public, and also to install a continuous emissions monitoring system.[13] But there would be more turmoil: Kearny officials continued to press facility owners and the NJDEP over the strong odors coming from the site, eventually suing the New Jersey Sports and Exposition Authority (NJSEA, the new name of the entity formerly known as the Meadowlands Commission) and NJDEP for access to their correspondence regarding the matter. In May, in a surprise ruling, Judge Jeffrey Jablonski issued a temporary injunction against the NJSEA, effectively closing the landfill. While the NJSEA claimed that they would suffer tremendous financial harm from any interruption in service, Judge Jablonski wrote: "The relief granted here is a temporary restraining order. This court finds . . . that immediate and irreparable harm will occur in the absence of the relief requested (for the continuous and unabated discharge of hydrogen sulfide), that the claim is based on a settled legal right that the Plaintiff (that seeks to protect the health of its citizens) will suffer greater harm if the condition is not mitigated nor abated, than the protection of the solely financial interest of the Defendant [NJSEA] in the site."[14]

Slowly failing facilities, well-intentioned oversight rife with loopholes, and disposal sites unable to handle all of the materials actually needing disposal—though the episodes outlined here are specific to New Jersey, they are representative of similar challenges confronting waste management elsewhere in the United States and even other parts of the world. A complicated set of issues faces everyone involved in the waste management community, ranging from the instability and unpredictability of markets for recyclable materials and other waste-derived products, to dwindling public finances; from advances in disposal technologies linked to high-tech startups, to large national and multinational corporations displacing mom-and-pop firms. More fundamentally, the waste management community faces a distinct political challenge: how to navigate the reality that absent a highly visible and disruptive crisis in waste disposal—something unlikely to occur in New Jersey anytime soon—neither the general public nor policymakers will demand improvements to New Jersey's system of waste management. While the episodes outlined above are in some instances shocking, all are essentially local concerns about odors, illegal dumping, or high spot prices for recycling service. We are not (yet) facing a statewide crisis in waste disposal capacity of the type that characterized the early and mid-1980s.

This final chapter offers a path forward, making a case for capitalizing on New Jersey's experience with trash while suggesting new policy and infrastructure approaches to waste management that could transform the state into a global leader in innovative collection, processing, and remanufacturing. Running counter to "zero-waste" hype and the pipe dream of eliminating waste entirely, this book concludes by acknowledging the reality that as humans, we will always be producing waste and requiring new strategies for handling it. I take the position that the environment, society, and economy of New Jersey will be much better off by embracing garbage as a building block for a sustainable future.

Opportunities and Obstacles on the Path Forward

New Jersey can build a prosperous future by capitalizing on existing strengths, expertise, and legal frameworks. This may be true of a number of industries found in the Garden State but is surely so for waste management. New Jersey can become a national and global leader in materials recovery, processing, and remanufacturing, and the foundations to achieve this goal—the legacy of the people, institutions, and problems examined in this book—are already in place. There is already a competitive marketplace for waste management services, including many businesses, large and small, public and private, that already collect and process waste materials (including materials that other jurisdictions have long ignored). There is historically a tradition of realistic, collaborative approaches to public oversight and an observable public commitment to supporting recycling programs. There is a robust collection of road, rail, and port infrastructures serving the state alongside an impressive industrial heritage. Finally, New Jersey has an enviable geographic location at the heart of both national and international flows of commerce that would allow innovative new materials derived from waste to reach markets around the globe.

Eliminate the Premium to Dispose Wastes in New Jersey

New Jersey lies at the center of the most populous region of the United States. Yet, as this book has shown, we continue to export our wastes, in increasing volumes, to disposal facilities and especially landfills elsewhere. This may be expedient but from an ethical perspective it is a dereliction of responsibility for cleaning up after ourselves and our guests. But sending wastes out of state appears cost effective, particularly when comparing disposal prices. According to a 2018 study completed by the Atlantic County Utilities Authority (ACUA), tipping fees in New Jersey ranged from $59.54 per ton at the Cumberland County Improvement Authority landfill complex to $130.55 per ton at the Covanta Essex County (Newark) WTE plant, with varying rates in between depending on any number of factors including facility type (landfill, WTE, transfer station, etc.) and the degree to which county government was able to reimpose some form of waste flow control after the *Atlantic Coast* decision in the late 1990s.[15] The average tipping fee in New Jersey was $79.73, compared with averages in Pennsylvania, New York, Maryland, Ohio, and Virginia—all places which have received shipments of New Jersey's wastes—of $66.61, $62.83, $68.69, $44.75, and $52.50, respectively.[16]

The apparent premium for the privilege of disposing wastes in New Jersey is clearly one legacy of the challenges with siting, building, and operating disposal sites in this most densely populated state. However, such a premium is far from permanent, and could be eliminated entirely through any number of policy measures, from a subsidy of tipping fees to some sort of export tax on wastes being shipped elsewhere. Consider the following example: from our analysis of the available data,[17] in 2015 New Jersey sent 11.25 percent of all municipal solid waste out of state. Assuming that all of that material, about 1.3 million tons, was *regular* waste, and the cost of disposal was simply the next highest tipping fee from the list provided above ($68.69 per ton), waste disposers in New Jersey saved $11.04 per ton by shipping waste elsewhere. But what if the state government had instead paid haulers the difference between New Jersey's average tip fee and the next highest tip fee, and all those sending waste out of state chose a New Jersey disposal facility instead? It would have cost the state about $14 million to equalize the prices. But if all 1.3 million tons of out-of-state waste had been disposed of at the average in-state rate, waste disposal facilities in New Jersey would have collected an additional $87 million.

From an in-state perspective, subsidizing the average tip fee could produce considerable additional value for disposal facilities in the state—money which could be used to revamp, repair, or clean up operations.[18] One logical conclusion of this thought experiment is to imagine what might happen if tip fees were subsidized to be the lowest in the northeast/mid-Atlantic region—how much additional revenue could disposal facilities generate if tip fees in New Jersey were the lowest in the region? What could we use that money for? Practically speaking, there is likely no need for New Jersey to have the absolute lowest tip fee in the region, because there is most definitely an optimization model waiting to be created that identifies the ideal rate to attract additional disposal while balancing various factors like transportation time, fuel costs, labor, and so forth. Around the globe, there have been examples of countries *importing* (on the other country's dime) wastes to keep disposal facilities and especially WTE plants running at full capacity.[19] Why not also New Jersey?

Invest in New Disposal and Processing Facilities

Should a subsidy on New Jersey's tipping fees be imposed tomorrow and waste haulers along the eastern seaboard decide to bring their trash to New Jersey, there would be tremendous problems. As discussed above, it is clear that most of New Jersey's disposal facilities are already approaching the end of their working lives or perhaps have even passed that point. At the moment, it is not entirely clear that disposal facilities in the state could handle even the additional volume of all in-state wastes. Thus, additional disposal facilities are needed in New Jersey, even to handle future flows of in-state waste, let alone facilitate a strategy of importing wastes from around the region. As this book has demonstrated, it is difficult if not impossible to site most types of industrial facility or environmental infrastructure in this state. However, in the same way that the premium for waste disposal in New Jersey is not a fact of nature, but rather the product of decades of policy and economic decisions, challenges with facility siting, operation, and upkeep can perhaps be overcome with careful planning and design.[20]

First, New Jersey should not pursue a strategy of large regional landfills, like "Edgeboro," or the Keegan Landfill, or of the type that dominate disposal in New York, Pennsylvania, or Ohio. At the same time, a realistic acceptance of landfills must be reached. We will, as humans, always need landfills because there will be some fraction of the waste stream that cannot be recycled or reduced any further, and New Jersey must carefully preserve existing landfill capacity. As Tom Marturano, former NJSEA director of solid waste noted, landfills are absolutely necessary to handle some odd materials:

TOM MARTURANO: Suppose there's some virus that comes along, wipes out the—all the chickens in New Jersey, of which I forget the number. It's an astounding number of chickens in New Jersey and something comes along and wipes them out . . . Because you're not going to be shipping that out to a state that has chickens. They're not going to allow you. They're just not going to allow you. So you better have a way of getting rid of that. So DEP actually does have a plan for that type of pandemic for livestock. The other thing people don't take into account is what about when big animals die? For example, my landfill, I'd bet you we've taken, at this point, eight whales.

JORDAN P. HOWELL: Whales?

TOM MARTURANO: Sure. Whales get killed all the time. Sure . . . They get nailed on the bow of a tanker ship [and] they're literally dragged at the front of the ship, because the ship's moving with that much momentum, until it gets to shore and then when the ship stops, the whale keeps going and crushes the dock. But now you have this 40, 50-ton animal that you need to get rid of. So there's a guy who you pay. You pay him 10 grand a day and he will come and he'll sling your whale out and he'll cut it up into pieces. But then he says, "Okay, where would you like me to bring it?" And that's when I get the call from the Army Corps [of Engineers] or the State Police, "We have a whale we want to bring you." And typically, what we'll do is we'll—if we know about it coming, and we always do, we dig a big hole and then when they come, we dump the whale in the hole and bury it with garbage. But I'll tell you a funny—

JORDAN P. HOWELL: Wait, I'm sorry. What was your reaction the first time to that phone call?

TOM MARTURANO: Well, the first time it happened, I got a call from the state police at 4:30 on a Friday afternoon. I was convinced it was a joke. A what? You know? He explained the whole thing and I—he said—so I said, "What about the other alternatives?" He says, "Well if you bury it at the beach, it pops back up." It gets bloated. The thing will pop back up. It's a problem. Plus it'll stink to high heaven. But what if you take it out to shore? First of all, scavengers are going to come. You're going to attract a lot of sharks. Some of it's going to wash its way back into the Jersey Shore. We wouldn't want that to happen. So what do you do with it? We got to get rid of it on land somehow. So we've picked you.[21]

As this, and many other waste industry anecdotes about spoiled beer, improperly labeled products, rancid garlic, and who know what else illustrate, landfills are still necessary even in the most advanced waste management systems. New Jersey must be prepared to handle its own whales, if nothing else, but more significantly be able to safely dispose of the fraction of the waste stream that cannot be recycled or reduced any further.

Having said that, landfills do not represent the future for New Jersey. Instead, the state should revisit plans for a network of WTE facilities and simultaneously consider co-siting anaerobic digesters to process food and organic wastes. Contemporary WTE is light-years ahead of old-style incinerators and even the first generation of WTE facilities rolled out in the United States in the 1970s and 1980s.[22] Modern WTE is clean, efficient, and a reliable source of baseload electric power or industrial steam. In 2019 as New Jersey aims to move toward greater reliance on wind and solar sources of electricity,[23] a network of new, efficient WTE plants can provide the necessary backup for these sometimes intermittent power producers.

One of the challenges facing WTE in the United States is a legacy of poor operations. Historically, incinerators and WTE plants sometimes operated outside of intended ranges, contributing to toxic emissions. But consider today that many of the ostensibly greenest countries on earth—Denmark, Sweden, Germany, Japan, and Switzerland, among others—have embraced WTE as their primary means of waste disposal. For example Denmark, roughly the size of Maryland and with a national population smaller than the state population of New Jersey, has about half as many WTE plants as there are in the entire United States. In these countries, emissions are closely monitored, and in virtually all instances, WTE facilities are connected to "district heating" networks that provide heating and cooling to nearby buildings. In another example of Denmark's embrace of WTE—as well as the safety of modern WTE plants—a recently opened plant in Copenhagen was designed not only as a hyperefficient waste disposal facility but also a recreational opportunity. Amager Bakke hosts a synthetic ski slope, open year-round, running from the top of its stack to ground level.

New-school WTE burns hotter and cleaner than the incinerator facilities of the past, but that does not mean that burning everything is always the ideal use of waste as a raw material. For example, organic wastes, which typically include food wastes from processing, retail, food service, and personal end use as well as many agricultural wastes and green wastes like leaves, grass, and clippings might have too much water content to burn efficiently in WTE. In the past many (including me) had advocated for classic styles of composting to recover the valuable organic material from this waste stream.[24] However it is not clear that New Jersey has the geographic space to host numerous large-scale composting operations, nor the agricultural, horticultural, and landscape industry markets to absorb huge quantities of finished compost. Instead of recreating many of the economic problems of recyclables by trying to force a market for compost, it would be better to invest in a network of anaerobic digesters that can rapidly biodegrade organic waste into biogas (similar to methane) and other byproducts. Biogas can be collected and used like methane, as a fuel for generating electricity or processed into a fuel for transportation. The other byproducts include solid and liquid components that can be used like fertilizers for farming. Given the large number of institutional organic waste producers in New Jersey—hospitals, universities, schools, malls, sports and event arenas, food processors, etc.—there is already considerable, concentrated production of food and organic waste that could be collected easily and taken to digester facilities. A network of digesters around the state could contribute to baseload electricity production, offer support to the remaining New Jersey farms, and help combat climate change by keeping decomposing, methane-producing organic materials out of landfills.

Second, New Jersey should aim to become the state with the ability to recycle or process *anything* into useful new material. That is to say, New Jersey should look beyond conventional recyclable materials and focus on processing unconventional or niche materials. There are already a few such operations in the state: for example, Bayshore Recycling has become a successful processor of contaminated soils; and Foam Cycle has partnered with a few of New Jersey's counties to process the expanded polystyrene foam (EPS) that serves as packing material for appliances and electronics. Trenton is home to Terracycle, an innovative processing firm embracing the mission (and social media tag) of #RecycleEverything.[25] The volumes of construction and demolition debris currently flowing into specialized landfills represent potential building materials for new or renovated housing.

Undoubtedly, there are other areas in which the state's waste management community has the knowledge—but maybe not the financial investment or market support—to process materials that would otherwise clog up remaining landfill space. Capitalizing on this expertise to become a national leader in materials processing need not rely on emerging high-technology, either—though we should also aim to support innovations in materials processing at New Jersey universities and startup firms. Is it so far-fetched to imagine New Jersey becoming a national or global center for processing electronic wastes? Currently, "e-waste" processing is an ecological and social disaster unfolding in some of the poorest places on earth, where exposure to highly toxic chemicals is a daily occurrence due to primitive processing techniques.[26] Could we not capitalize on the ingenuity, labor base, and industrial knowledge of this state to create a better, safer, more effective way? What if New Jersey companies became leaders in processing discarded textiles into new fabrics? Or champions of other types of disassembly, from old appliances to furniture and everything in between?

New Jersey's existing transportation infrastructure would facilitate the success of these new disposal and processing facilities. We are at the crossroads of the U.S. East Coast and a major shipping hub. There are already extensive road, rail, and port facilities here, connecting not only communities within New Jersey, but the Garden State to the rest of the United States and even the planet. New facilities should be sited to receive wastes especially from rail- and water-borne modes of transportation, and we should support rail and water export terminals in other states to facilitate import of their wastes. Facility siting—the bête noire of New Jersey's waste management history—simply must capitalize on existing and disused industrial sites. One major selling point of WTE plants, anaerobic facilities (particularly those for urban and institutional organic wastes), and materials processing plants is that they are very much **not** spatially extensive large landfills; they are compact industrial facilities. Locating new disposal and processing facilities in existing industrial areas is therefore not only feasible, as with a brownfields redevelopment,[27] but could pay environmental and economic dividends. To offer another example from Denmark, the famous industrial park in Kalundborg incorporates waste processing and WTE as an important role in the industrial ecosystem connecting several different production facilities including an oil refinery, pharmaceuticals manufacturer, water utility, and gypsum board manufacturer among others. All of these plants exchange energy, water, and raw materials in a network with one another.[28]

Careful attention to engineering and design could tie new WTE and anaerobic digestion facilities into the electrical and steam systems of New Jersey's industrial facilities in a similar fashion, simultaneously making the state's economy more competitive and environmentally sustainable. Proposals for the development of these type of industrial ecosystems and "microgrids" are already being considered in Trenton and also Camden, linking together various facilities so that they may be able to sustain operations in the event of a widescale power outage or other crisis.[29] Alternatively, new waste disposal facilities and especially WTE and anaerobic plants could supply free steam and electricity service to the many dense urban neighborhoods and towns that can be found across the state, lowering for host communities the state's painfully high costs of living. These suggestions, while ambitious, are not fantasy—they already exist in some form or format around the world. To be implemented in New Jersey, however, would require not only brave public support for new and rehabilitated infrastructure but other economic policy shifts.

Create and Support Better Markets for Postconsumer Materials

Considerable time, energy, and money has been expended on recycling in New Jersey and indeed in most places in the United States. Born in response to concerns over ecological problems but also as a strategy for extending the life of landfills, curbside residential recycling surely came from a place of good intentions. Extensive research has since demonstrated, however, that even before the collapse of markets for recyclable materials linked to a change in Chinese policy, U.S. residential recycling had become in many ways an expenditure of nervous energy—an opportunity to *do something* about the environment without actually doing anything much at all, at increasingly high cost.[30] The underlying reason for this failure is because the supply chain of moving recyclables collected from homes and businesses, to processors, to end users making new products is far less consistent than the types of flows typical of other commodity materials. Why? Two truths about the market for postconsumer materials are worth noting here:

1 Markets (i.e., prices) for recyclable materials are cyclical. Recycling processing firms already operate on thin margins; downturns in markets for recyclables not only hurt existing firms' ability to function but also (likely) discourage new firms from entering the market and (perhaps) discourage broader investment in the industry.

2 Recyclables are a substitutable commodity. That means, for example, recycled plastic has to compete with new plastic when a firm is looking to manufacture packaging; recycled paper with new paper; recycled cardboard with new cardboard, and so on. Therefore, and first, recyclables' prices have to react to the prices of other, substitutable commodities; and second, there is likely a "premium" in the prices of recyclables over new materials due to the processing costs that makes them less attractive as inputs to production.

Variabilities in price, availability, and quality mean recyclable materials are typically much riskier for companies to work with. Efforts to reform markets for recycled materials must focus on smoothing out these variabilities as much as possible if recycling is to achieve its maximum economic and ecological potential. It is no longer effective to try and make moral arguments for recycling—that people should separate materials into different piles because "it's the right thing to do" or because it "helps the environment." Markets for recyclables in the United States will only be resuscitated when the underlying material becomes so valuable that people take care to recycle because it benefits their pocketbook, or companies utilize recycled materials in their products because they are cheaper than other alternatives and consistently available when needed. Smart policy can produce these outcomes. New Jersey government (though ideally, federal government) should directly support domestic markets for recycled material in order to make the costs, access, and quality competitive with new material.

One way of doing that might be to look to an approach once taken in the energy sector. In 1978 the federal government passed the Public Utilities Regulatory Policies Act (PURPA) which established a minimum price for electricity produced by new "qualifying facility" types of companies and new sources of electricity.[31] In effect, PURPA *required* established utility companies to purchase power from qualifying facilities at an "avoided cost" rate—in other words, the price the utility company would have (likely) paid to generate the same amount of electricity themselves. With this approach, the qualifying facilities were guaranteed both a buyer for their product and an incentive to maximize the difference between the "avoided cost" rate and their actual cost for producing the power, by improving efficiency as much as possible. Imagine a similar structure for the recyclables industry, where taxes are imposed on new materials to make them equivalent in price to recycled materials coming out of "qualified facilities." The tax revenues could flow back to recycling collectors and processors (or even consumers, if a container refund law were implemented[32]). In that scenario, a manufacturer (for example) of plastic bottles would have to pay the same amount for either new or recycled plastic—they could still choose—but with the key difference that there would be tremendous profit potential for new firms to enter both the recycling collection and processing business. Over time, more plastic would be recovered, increasing the supply of recycled plastic, and perhaps lowering prices. The strategy could be useful in supporting markets for currently low-value recyclable materials by making them profitable to collect, process, and reuse as inputs to new production. Since recyclable materials are cyclical products, with implications for business cycles, they currently appear to attract less investment attention because returns are less consistent.[33] With an "avoided cost" approach, a predictable price for recycled materials could be established, and perhaps even linked to some sort of regional, national, or global index. Predictable prices and sales volumes should make it easier for the recycling industry to attract capital investment of all types.

Similarly, another way of supporting markets for recycled materials and in particular, smoothing volatility in pricing and material supply could be the development of derivative products—options and futures contracts in particular—for postconsumer materials. Financial instruments for recyclables have been pondered before,[34] and indeed one exchange for paper pulp and corrugated cardboard has emerged in Norway, NOREXECO. But in the era of financial engineering in which we currently live, it is perhaps surprising that derivatives for recycling are not more widely available. As researcher Matthias Herskind argues,

> Federal, state and local governments need to realize that they must not only permit, but support a profitable transition for waste operators. Instead of making waste operators "promise" recycling goals to acquire attractive disposal contracts, the recycling contracts themselves should (by all actionable means) be made attractive . . .
>
> [One] step to improving the attractiveness profile on a capital investment-basis is to continue building liquid secondary markets with derivatives for hedging commodity exposure. The current market size for commodities like recycled mixed paper is enormous, but the financial sophistication is relatively low. A secondary market with liquid trading of futures for [materials like]

recycled pulp would increase stability through other means and allow producers throughout the value chain to allocate the commodity price risk to willing risk-takers. This will require an exchange with price-quotes and rules for the commodities traded ... Hedging opportunities with proper liquidity are only available on widely-traded virgin commodities, not their recycled cousins. Building a liquid secondary market with live quotes and derivatives will enhance stability.[35]

The NOREXECO exchange, for example, describes itself as "a regulated commodity exchange specialised for the global pulp and paper industry" located in Norway.[36] NOREXECO operates with the same commodities trading principles as an entity like the well-known Chicago Mercantile Exchange, except that instead of trading contracts for pork bellies or canola oil, NOREXECO participants deal in different types of paper pulp and corrugated cardboard.

Derivatives like commodities futures contracts are beneficial for both buyers and sellers of the underlying commodity. If parties think the price of a commodity is going to change in the coming months, a futures contract allows a price to be locked in now, benefitting buyers or sellers depending on the purpose of buying the contract (i.e., going "long" or "short").[37] Futures markets can also give end users of a particular commodity indication about changes in supply, which is useful for firms planning to use that commodity in their productive activities because potential shortages (or surpluses) would be reflected by changing prices. Commodities futures will tie together different geographic regions, creating a global or near-global market for a single commodity with much deeper overall liquidity.[38] In this scenario, one might imagine a three-month futures contract for recovered aluminum, textiles, EPS, or e-waste components—maybe having just been processed in one of New Jersey's new industrial and waste management ecosystems imagined earlier—being quoted to buyers in Michigan, Mexico, or Mongolia for shipment from the Port of Camden or the Port Newark-Elizabeth Marine Terminal.

Market development is nothing new in the world of recycling, but many earlier policy efforts focused on direct subsidy for particular plants and projects, public education and outreach efforts, and support for narrow, niche end market products emphasizing the reuse or recycling of one particular type of material. These types of support have perhaps been useful in raising awareness about recycling or getting processing facilities up and running, but less useful in creating broad, liquid markets for recycled materials that see them reliably integrated back into manufacturing processes. New Jersey could have the most wonderful recycling processing in the world, but with unreliable end markets, such projects are white elephants. By focusing on measures for attracting investment capital and building liquidity into recycled commodities markets, the solutions suggested here emphasize ways of making broad categories of recycled materials more profitable and reliable in the supply chain. If history is any guide, this will attract new firms and investment to the sector, eager to cash in. New Jersey could, and should, become the epicenter for financial dimensions of waste management as well; an investment which will, over time, reap considerable economic and ecological rewards.

Conclusion: From Garden State to Garbage State?

All of the ideas and suggestions in this chapter center on the notion of collecting, importing, processing, and disposing more waste in New Jersey. But this book has examined the considerable anguish surrounding management of even smaller volumes of waste over the past fifty years. And so the suggestions made here inevitably beg the question: why on earth should we want to dispose of more garbage in New Jersey?

Nothing has been implemented quickly or easily in New Jersey's waste history. Indeed, this book has shown the challenges with designing, planning, and operating waste management infrastructure here. Looking back, we have moved through many phases of waste management in New Jersey, starting from a place with little centralized oversight (or control) through to a few years of a unitary, comprehensively planned system that ultimately faltered when the highest court in the land struck down core concepts and tactics of that system. New Jersey's waste history is characterized by state government taking a leading role, and then working to convince and cajole both other levels of government and private industry alike into coming along. Even as technologies (landfilling, WTE, recycling, etc.), financing issues, industry oversight (utilities regulation, attempts to root out organized crime), and public attitudes changed, state agitation and insistence on handling New Jersey's wastes in the places where it is produced has remained consistent. Throughout this time, politically and perhaps also practically, it would have been far easier to simply export waste flows to landfills elsewhere than it would have been to undertake the effort of siting and financing new facilities here. Yet ultimately the attempt at self-sufficiency led by the state government must be commended.

And it is in that spirit that these suggestions are made. Still, critics might say that the approaches outlined in this chapter do not necessarily solve the problem of waste anyway; and, I would agree on that point. But the most important conclusion I reach in working on this book, and I hope that readers will reach as well, is that waste is not a problem to ever be fully solved. The archaeological record shows that for as long as humans have been around, we have been producing wastes and struggling to find a fully satisfying solution for that material, whether a burning pile of refuse or a dump on the outskirts of town. A strategy of "out of sight, out of mind" is no longer suited to the ecological challenges we face in the world today nor does it match our human intellect and collective capability to do something better. Aside from the contention that exporting waste is a considerable outflow of cash—typically the dollars of businesses and residents—to firms in other states, an exporting strategy is also a dereliction of environmental responsibility. We have to recognize that if we in New Jersey are choosing to consume, we also have the obligation to clean up the ensuing wastes. I propose in this chapter some strategies for making that cleanup more effective and perhaps financially attractive and that build on the legacy of the state's waste management journey to this point.

New Jersey's systems for solid waste management are ripe for investment and reinvention, and offer a tantalizing economic, ecological, and social opportunity. Currently, we have a once-in-a-generation opportunity to capitalize on New Jersey's experience with trash and transform the state into a global leader in innovative waste collection, processing, and remanufacturing. The facilities we currently use here are reaching the end of their lives. Why not make the next wave of waste management infrastructure an example for the rest of the planet? We can choose to embrace the twin historical realities of being both a regional epicenter for trash disposal and an innovator in waste management policy to become the global leader in sustainable waste management, or let someone else take the lead. If we do not make these types of changes in New Jersey, someone else surely will—there is too much money at stake.

Additional work is needed to advance this vision. Instead of chasing narrow industrial projects or courting tech giants with tax giveaways, public and private investment ought to focus on new waste processing and manufacturing technologies. Instead of killing innovative technologies by a thousand regulatory cuts, adopt a catholic approach to incubating new firms that deal in waste collection, processing, brokering, and financing. A prosperous garbage future for New Jersey would require establishing supports for the markets for waste management goods and services as well as streamlining data collection, permitting and regulatory processes. Most important, however, would be a shift in mindset that reorients our thinking about waste management from a problem to an environmental and economic opportunity.

Acknowledgments

If you haven't written one, you might not know much about the process of publishing a book. Rest assured that there are many people involved beyond the author, each with an important role. I'd like to acknowledge the support of several institutions who provided funding for the research underlying this book. In particular, this book is one of the main products of my project *Ecological Identity and Solid Waste Management,* funded by the Science, Technology and Society Program at the U.S. National Science Foundation (award number 1733924). Special thanks to the interlibrary loan and circulation staff at the Campbell Library at Rowan University for facilitating my voluminous requests for materials, and to all the libraries that were willing to engage in resource sharing. It's as important as ever that we as a society create, organize, maintain, and make available our public records in their entirety. The views presented in this book are my own and do not necessarily represent those of any office at the National Science Foundation or any office at Rowan University.

Thanks to the many solid waste management professionals who participated in interviews with me: Gary Sondermeyer, Joe Davis, Guy Watson, Dr. Marwan Sadat, Wayne DeFeo, George Tyler, John Weingart, Jack Sworaski, Tom Marturano, Larry Gindoff, Rick Dovey, John Waffenschmidt, Valerie Montecalvo, Reenee Casapulla, William Pikolycky, Charles Norkis, Mitchell Kizner, and John Wohlrab. Thanks also to Carlton Dudley and Connor Loftus of the New Jersey Department of Environmental Protection for helping to track down and transmit to me nearly thirty years of waste disposal and recycling data for the many towns and counties of New Jersey.

A special thanks to Gary Sondermeyer: Gary is a tireless advocate of the importance of proper waste management and a bottomless well of information about solid waste management and the history of waste management in New Jersey. Additionally, I have never seen Gary without a smile on his face or a kind word. This book could not have been written without his help. Special thanks also to Martin Melosi for "talking trash" with me at several meetings of the American Society for Environmental History over the course of this project.

Thanks to those who reviewed early draft chapters and materials: Edward Lloyd and Mark Lohbauer; my friend and colleague Dr. Zachary Christman; tennis associate Tom "Hammer" Hurley, Esq.; and the family members below. I'd like to thank Peter Mickulas, executive editor for History and Sociology at Rutgers University Press, for his time, insights, and energy in shepherding this book through the publication process.

Finally, to the friends and family who have provided moral, material, and intellectual support over the years that this process has unfolded—none of this would have been possible without you! Thanks to my wife, Amanda; my daughter, Paloma; my parents, Pam and Tom; our family members; and our beloved cat, Ms. Marilyn (RIP).

Portions of chapter 3 previously appeared in the journal *Heliyon* as "New Jersey's Solid Waste and Recycling Tonnage Data: Retrospect and Prospect." The full citation for the article is: Jordan P. Howell, Katherine Schmidt, Brooke Iacone, Giavanni Rizzo, and Christina Parrilla, "New Jersey's Solid Waste and Recycling Tonnage Data: Retrospect and Prospect," *Heliyon* 5, no. 8 (2019), https://doi.org/10.1016/j.heliyon.2019.e02313.

Jordan P. Howell, PhD, MBA
May 13, 2022
Rowan University

Notes

Chapter 1   Introduction

1 "The Sopranos," *The Sopranos*, HBO, January 10, 1999.
2 "A Big Piece of Garbage," *Futurama*, FOX, May 11, 1999.
3 The idea that the New Jersey Turnpike conveys the very worst image possible of the state, environmentally, is an oft-referenced trope in writing about New Jersey. See also Patricia Ard, "Garbage in the Garden State: A Trash Museum Confronts New Jersey's Image," *The Public Historian* 27, no. 3 (2005): 57–66; Neil Maher, ed., *New Jersey's Environments: Past, Present, and Future* (New Brunswick, NJ: Rutgers University Press, 2006); Thomas Belton, *Protecting New Jersey's Environment: From Cancer Alley to the New Garden State* (New Brunswick, NJ: Rivergate Books, 2011).
4 Jennifer Brown, "N.J. Beachgoers Leave Nearly a Half-Million Pieces of Trash Behind," *The Star-Ledger*, April 19, 2011, https://www.nj.com/news/index.ssf/2011/04/nj_beachgoers_leave_nearly_a_h.html.
5 See James W. Hughes and Joseph J. Seneca, *New Jersey's Postsuburban Economy* (New Brunswick, NJ: Rutgers University Press, 2015).
6 Bureau of Economic Analysis, 2018, "New Jersey," United States Dept. of Commerce, accessed July 31, 2018, https://www.bea.gov/regional/bearfacts/pdf.cfm?fips=34000&areatype=STATE&geotype=3&mod=articla_inline.
7 See Maxine N. Lurie, ed., *A New Jersey Anthology* (New Brunswick, NJ: Rutgers University Press, 2010); and Maxine N. Lurie and Richard F. Veit, eds., *New Jersey: A History of the Garden State* (New Brunswick, NJ: Rutgers University Press, 2012).
8 For discussion on the philosophical dimensions of waste, see: Joshua Reno, "What is Waste?" *Worldwide Waste: Journal of Interdisciplinary Studies* 1, no. 1 (2018), http://doi.org/10.5334/wwwj.9; Mary Douglas, *Purity and Danger: An Analysis of Concepts of Pollution and Taboo* (New York: Praeger, 1966); Susan Strasser, *Waste and Want: A Social History of Trash* (New York: Metropolitan Books, 1999).
9 United States Environmental Protection Agency, "National Overview: Facts and Figures on Materials, Wastes, and Recycling," accessed July 31, 2018, https://www

.epa.gov/facts-and-figures-about-materials-waste-and-recycling/national-overview
-facts-and-figures-materials#NationalPicture.

10 Great controversy surrounds the accounting of industrial wastes in the United
States; see Jonathan S. Krones, "Accounting for Non-Hazardous Industrial Waste
in the United States" (PhD diss., Massachusetts Institute of Technology, 2016).

11 Per the NJDEP: "A solid waste is defined in New Jersey's Solid Waste Regulations
as any garbage, refuse, sludge, or any other waste material. Certain exemptions are
made under the regulatory definition, including source separated food wastes
collected for livestock feed, certain recyclable materials, spent sulfuric acid used to
produce virgin sulfuric acid, and certain dredged materials, among others.
N.J.A.C. 7:26-1.6 provides additional details. The definition of solid waste
includes a wide variety of materials that have served or can no longer serve their
original intended use that are discarded, intended to be discarded, accumulated in
lieu of being discarded, or burned for energy recovery. Solid waste includes
residential, commercial, and institutional solid waste generated within a community, which is termed municipal solid waste, or MSW. Solid waste also includes
items such as construction and demolition waste and bulky waste including
appliances and furniture. Certain solid wastes are classified as hazardous wastes,
and are subject to specific management requirements. Certain other materials are
exempted from the solid waste definition pursuant to New Jersey's Recycling
Regulations, for example tree branches, limbs, trunks, brush and wood chips that
will be received, stored, or processed in accordance with the regulations." New
Jersey Dept. Environmental Protection, 2017, *Solid Waste and Recycling Environmental Trends Report*, NJDEP Division of Science, Research, and Environmental
Health, accessed July 31, 2018, 1, https://www.nj.gov/dep/dsr/trends/pdfs
/solidwaste.pdf.

12 New Jersey data from *Solid Waste and Recycling Environmental Trends Report*, 2;
EPA data from "National Overview: Facts and Figures on Materials, Wastes, and
Recycling."

13 Canada data from Government of Canada, 2016, "Solid Waste Diversion and
Disposal," accessed July 31, 2018, https://www.canada.ca/en/environment-climate
-change/services/environmental-indicators/solid-waste-diversion-disposal.html;
EU data from Eurostat, 2018, "Municipal Waste Statistics," accessed July 31, 2018,
http://ec.europa.eu/eurostat/statistics-explained/index.php/Municipal_waste
_statistics#Municipal_waste_generation; Japan data from Christine Yolin, 2015,
*Waste Management and Recycling in Japan: Opportunities for European Companies
(SMEs focus)*, EU-Japan Centre for Industrial Cooperation, accessed July 31, 2018,
29, https://www.eu-japan.eu/sites/default/files/publications/docs/waste
_management_recycling_japan.pdf.

14 The work of these three scholars in particular has shaped my own approach
toward waste management research. See Martin V. Melosi, *Garbage in the Cities:
Refuse, Reform, and the Environment*, rev. ed. (Pittsburgh, PA: University of
Pittsburgh Press, 2005); Martin V. Melosi, *The Sanitary City: Urban Infrastructure
in America from Colonial Times to the Present* (Baltimore: Johns Hopkins
University Press, 2000); Samantha MacBride, *Recycling Reconsidered: The Present
Failure and Future Promise of Environmental Action in the United States* (Cam-
bridge, MA: The MIT Press, 2012); and Richard C. Porter, *The Economics of Waste*
(Washington, DC: Resources for the Future Press, 2002).

15 For some illustrations of how this tension played out in Hawaii, see Jordan P. Howell, "The Fate of Waste in Hawaii: Technology Assessment and Solid Waste Planning in Hawaii, 1968–78," *Singapore Journal of Tropical Geography* 36 (2015): 67–82; Jordan P. Howell, "'Modes of Governing' and Solid Waste Management in Maui, Hawaii, USA," *Environment and Planning A: Economy and Space* 47 (2015): 2153–2169; Jordan P. Howell, "Alternative Waste Solutions for the Pacific Region: Learning from the Hawai'i Experience," *East-West Center AsiaPacific Issues* 121 (2015): 1–8; and Jordan P. Howell, "Waste Governance and Ecological Identity in Maui, Hawaii, USA," *Geoforum* 79 (2017): 81–89.

16 See, Howell, "'Modes of Governing'" for an instance of private firms shaping the waste disposal systems of Maui, Hawaii.

Chapter 2   Origins of Waste Management Planning in New Jersey

Epigraph Source: Special Legislative Commission to Investigate Certain Problems Relating to Solid Waste Disposal, *Public Hearing before Special (Legislative) Commission to Investigate Certain Problems Related to Solid Waste Disposal (Constituted under SCR 24 of 1969)* (Trenton: New Jersey Legislature, 1969), vol. 4, p. 5–6.

1 George Tyler, interview by Jordan P. Howell, December 5, 2017.
2 The NJSEA (New Jersey Sports and Exposition Authority) is the current name of the former Hackensack Meadowlands Development Commission (HMDC). Thomas Marturano, interview by Jordan P. Howell, April 5, 2018.
3 Commission to Study the Problem of Solid Waste Disposal, *Public Hearing Before Commission to Study the Problem of Waste Disposal: (Created Under Provisions of Assembly Concurrent Resolution no. 36)* (Trenton: New Jersey Legislature, 1965), 1.
4 *Public Hearing Before Commission to Study the Problem of Waste Disposal*, 3–7.
5 New Jersey State Department of Health. State Sanitary Code. Chapter VIII, Refuse Disposal (Trenton, NJ: 1957), 1.
6 Commission to Study the Problem of Solid Waste Disposal, *Public Hearing Before Commission to Study the Problem of Waste Disposal*, 10.
7 New Jersey State Department of Health: Division of Clean Air and Water - Solid Waste Disposal Program, *The New Jersey Solid Waste Disposal Program* (Trenton, NJ: 1968).
8 New Jersey State Department of Health, *The New Jersey Solid Waste Disposal Program*, 2.
9 New Jersey State Department of Health, *The New Jersey Solid Waste Disposal Program*, 3.
10 New Jersey State Department of Health, *The New Jersey Solid Waste Disposal Program*, 3.
11 New Jersey State Department of Health, *The New Jersey Solid Waste Disposal Program*, 3–4.
12 New Jersey State Department of Health, *The New Jersey Solid Waste Disposal Program*, 5.
13 New Jersey State Department of Health, *The New Jersey Solid Waste Disposal Program*, 6.
14 Special Legislative Commission to Investigate Certain Problems Relating to Solid Waste Disposal, *Public Hearing before Special (Legislative) Commission*, vol. 2, 4.

15 Special Legislative Commission to Investigate Certain Problems Relating to Solid Waste Disposal, *Public Hearing before Special (Legislative) Commission*, vol. 2.

16 See, Robert Sullivan, *The Meadowlands: Wilderness Adventures at the Edge of a City* (New York: Scribner, 1998).

17 Special Legislative Commission to Investigate Certain Problems Relating to Solid Waste Disposal, *Public Hearing before Special (Legislative) Commission*, vol. 2, 28.

18 Special Legislative Commission to Investigate Certain Problems Relating to Solid Waste Disposal, *Public Hearing before Special (Legislative) Commission*, vol. 2, 32.

19 Special Legislative Commission to Investigate Certain Problems Relating to Solid Waste Disposal, *Public Hearing before Special (Legislative) Commission*, vol. 2, 47–48.

20 Special Legislative Commission to Investigate Certain Problems Relating to Solid Waste Disposal, *Public Hearing before Special (Legislative) Commission*, vol. 2, 37.

21 Special Legislative Commission to Investigate Certain Problems Relating to Solid Waste Disposal, *Public Hearing before Special (Legislative) Commission*, vol. 2, 35–36.

22 Special Legislative Commission to Investigate Certain Problems Relating to Solid Waste Disposal, *Public Hearing before Special (Legislative) Commission*, vol. 2, 50–51.

23 Special Legislative Commission to Investigate Certain Problems Relating to Solid Waste Disposal, *Public Hearing before Special (Legislative) Commission*, vol. 2, 31–32.

24 Special Legislative Commission to Investigate Certain Problems Relating to Solid Waste Disposal, *Public Hearing before Special (Legislative) Commission*, vol. 2, 33.

25 Special Legislative Commission to Investigate Certain Problems Relating to Solid Waste Disposal, *Public Hearing before Special (Legislative) Commission*, vol. 2, 37–38.

26 Special Legislative Commission to Investigate Certain Problems Relating to Solid Waste Disposal, *Public Hearing before Special (Legislative) Commission*, vol. 2, 41–42.

27 This section of the testimony transcript ends abruptly after Mr. McCurrie's response. Special Legislative Commission to Investigate Certain Problems Relating to Solid Waste Disposal, *Public Hearing before Special (Legislative) Commission*, vol. 2, 46.

28 Special Legislative Commission to Investigate Certain Problems Relating to Solid Waste Disposal, *Public Hearing before Special (Legislative) Commission*, vol. 2, 56–66.

29 Special Legislative Commission to Investigate Certain Problems Relating to Solid Waste Disposal, *Public Hearing before Special (Legislative) Commission*, vol. 2, 63.

30 Special Legislative Commission to Investigate Certain Problems Relating to Solid Waste Disposal, *Public Hearing before Special (Legislative) Commission*, vol. 2, 67–68.

31 Special Legislative Commission to Investigate Certain Problems Relating to Solid Waste Disposal, *Public Hearing before Special (Legislative) Commission*, vol. 2, 70–73.

32 Special Legislative Commission to Investigate Certain Problems Relating to Solid Waste Disposal, *Public Hearing before Special (Legislative) Commission*, vol. 2, 69.

33 Special Legislative Commission to Investigate Certain Problems Relating to Solid Waste Disposal, *Public Hearing before Special (Legislative) Commission*, vol. 2, 68.

34 Special Legislative Commission to Investigate Certain Problems Relating to Solid Waste Disposal, *Public Hearing before Special (Legislative) Commission*, vol. 4.

35 Special Legislative Commission to Investigate Certain Problems Relating to Solid Waste Disposal, *Public Hearing before Special (Legislative) Commission*, vol. 4, 2–3.

36 Special Legislative Commission to Investigate Certain Problems Relating to Solid Waste Disposal, *Public Hearing before Special (Legislative) Commission*, vol. 4, 5–6.

37 Special Legislative Commission to Investigate Certain Problems Relating to Solid Waste Disposal, *Public Hearing before Special (Legislative) Commission*, vol. 4, 26–27.

38 Special Legislative Commission to Investigate Certain Problems Relating to Solid Waste Disposal, *Public Hearing before Special (Legislative) Commission*, vol. 4, 31–32.

39 Monmouth County Planning Board, *Study & Plan of Refuse Collection and Disposal, Monmouth County* (Monmouth County Board of Chosen Freeholders, 1966); Frank Perrotta, *An Updated Study of Bergen County's Solid Waste Disposal System* (Hackensack, NJ: Bergen County Board of Chosen Freeholders, 1969); Camden County Planning Board, *Solid Waste, 1970* (Pennsauken, NJ: Camden County Planning Board, 1970); Mercer County Planning Board, *Mercer County, New Jersey Comprehensive Plan: Solid Waste Disposal Study and Plan*. (Philadelphia, PA: Day & Zimmermann, Inc., 1971); Township of Woodbridge, and New Jersey Department of Community Affairs, *Solid Waste Study of the Township of Woodbridge, New Jersey* (Trenton, NJ: New Jersey Department of Community Affairs, 1970).

40 See, for example, Mary McLean, *Planning for Solid Waste Management* (Chicago, IL: American Society of Planning Officials, 1971); or for the role of engineering societies and technical offices in developing solid waste plans see, Martin V. Melosi, *Garbage in the Cities: Refuse, Reform, and the Environment*, rev. ed. (Pittsburgh, PA: University of Pittsburgh Press, 2005); Jordan P. Howell, "Technology and Place: A Geography of Waste-to-Energy in the United States" (PhD diss., Michigan State University, 2013).

41 Camden County Planning Board, *Solid Waste, 1970*, preface.

42 Camden County Planning Board, *Solid Waste, 1970*, 7.

43 Camden County Planning Board, *Solid Waste, 1970*, 8.

44 Camden County Planning Board, *Solid Waste, 1970*, 1.

45 Planners Associates, Inc., *New Jersey State Solid Waste Management Plan* (Newark, NJ: Planners Associates Inc. and Bureau of Solid Waste Management, New Jersey Department of Environmental Protection, 1970).

46 Planners Associates, Inc., *New Jersey State Solid Waste Management Plan*, preface.

47 Planners Associates, Inc., *New Jersey State Solid Waste Management Plan*, preface.

48 Planners Associates, Inc., *New Jersey State Solid Waste Management Plan*, 1–2.

49 Planners Associates, Inc., *New Jersey State Solid Waste Management Plan*, 5.

50 Planners Associates, Inc., *New Jersey State Solid Waste Management Plan*, 7.

51 Planners Associates, Inc., *New Jersey State Solid Waste Management Plan*, 7.

52 Planners Associates, Inc., *New Jersey State Solid Waste Management Plan*, 8.

53 Planners Associates, Inc., *New Jersey State Solid Waste Management Plan*, 7.

54 Planners Associates, Inc., *New Jersey State Solid Waste Management Plan*, 52.

55 Planners Associates, Inc., *New Jersey State Solid Waste Management Plan*, 1.

56 Planners Associates, Inc., *New Jersey State Solid Waste Management Plan*, 13.

57 Planners Associates, Inc., *New Jersey State Solid Waste Management Plan*, 14.

58 Planners Associates, Inc., *New Jersey State Solid Waste Management Plan*, 61.

59 Planners Associates, Inc., *New Jersey State Solid Waste Management Plan*, 62.

60 The plan also briefly considers a third planning alternative, identified as the "railhaul strategy" which involves reliance on New Jersey's freight rail infrastructure to ship waste to unidentified disposal sites within or beyond state borders. This approach is poorly articulated in the plan, but is summarized as follows: "Of the less traditional methods of waste disposal available to county or group of municipalities, transferring the material to a railway and hauling it outside of the area presents an attractive possibility. The chief problem with this method has

thus far been political: most jurisdictions have been reluctant to receive solid wastes generated by other areas . . . For such a system to be feasible, a substantial portion of New Jersey's solid waste must be transferred and rail hauled to large sites for disposal . . . It does not appear that this system presents an immediate alternative to the State." Planners Associates, Inc., *New Jersey State Solid Waste Management Plan*, 91.

61 Planners Associates, Inc., *New Jersey State Solid Waste Management Plan*, 95.

62 Planners Associates, Inc., *New Jersey State Solid Waste Management Plan*, 96.

63 Planners Associates, Inc., *New Jersey State Solid Waste Management Plan*, 96.

64 Planners Associates, Inc., *New Jersey State Solid Waste Management Plan*, 96.

65 Planners Associates, Inc., *New Jersey State Solid Waste Management Plan*, 90.

66 Planners Associates, Inc., *New Jersey State Solid Waste Management Plan*, 92.

67 Planners Associates, Inc., *New Jersey State Solid Waste Management Plan*, 13.

68 Tocks Island Regional Advisory Council, and Candeub Fleissig and Associates, *Tocks Island Regional Interstate Solid Waste Management Study* (Stroudsburg, PA: Tocks Island Regional Advisory Council, 1969); Metropolitan Regional Council, *Solid Waste Management in the Region: An Evaluation and Report to Local Government Officials in the New York-New Jersey-Connecticut Metropolitan Area* (New York: Metropolitan Regional Council, 1971).

69 Tocks Island Regional Advisory Council, *Tocks Island Regional Interstate*, 3.

70 Tocks Island Regional Advisory Council, *Tocks Island Regional Interstate*, 32.

Chapter 3   Planning, Siting, Operating, and Financing Landfills

Epigraph Source: Cited in Walter H. Waggoner, "State Orders Tight Curbs on Outside Trash Haulers," *New York Times*, October 20, 1973, 67.

1 New Jersey Department of Community Affairs, *Community* (Trenton: New Jersey Department of Community Affairs, July 1971), 1.

2 New Jersey Department of Community Affairs, *Community*, 6.

3 New Jersey Department of Community Affairs, *Community*, 1.

4 New Jersey Department of Community Affairs, *Community*, 1, 6.

5 Division of Waste Management, New Jersey Department of Environmental Protection, *Solid Waste Management Plan: Draft Update, 1985–2000* (Trenton: New Jersey Department of Environmental Protection, 1985), 80.

6 Planners Associates, Inc., *New Jersey State Solid Waste Management Plan* (Newark, NJ: Planners Associates Inc. and Bureau of Solid Waste Management, New Jersey Department of Environmental Protection, 1970), 61.

7 R. Carson, *Silent Spring* (Boston: Houghton Mifflin, [1962] 2002); Adam Rome, *The Bulldozer in the Countryside: Suburban Sprawl and the Rise of American Environmentalism.* (Cambridge: Cambridge University Press, 2001).

8 William V. Musto and The County and Municipal Government Study Commission, *Solid Waste: A Coordinated Approach* (Trenton, NJ: County and Municipal Government Study Commission, 1972), 35.

9 Musto and Commission, *Solid Waste: A Coordinated Approach*, 5.

10 Dennis J. Krumholz, "Legal Issues in Solid Waste Management," *New Jersey Lawyer* (1983): 2, http://riker.com/print/publications/1983-legal-issues-in-solid-waste-management.

11 Per Krumholz "Legal Issues in Solid Waste Management," 8: "See, for example, *In the Matter of the Closure Order of April 23, 1979 issued to Henry Harris Landfill,*

(A-1787–80 T4. decided February 25, 1981) (landfill operations terminated because facility was placing waste at elevations in excess of those set forth in facility's approved engineering design); *State at New Jersey, Department of Environmental Protection v. Sanitary Landfill, Inc.*, (A-450-80T2, decided July 19, 1982) (landfill terminated because facility was operating at elevations in excess of those approved by the NJDEP); *Gloucester Environmental Management Services, Inc. v. State of New Jersey, Department at Environmental Protection* (A-550-78, A-4535-78) (landfill operations enjoined because facility was operating without approved engineering design and in an environmentally unsound manner)."

12 New Jersey State Commission of Investigation, *A Report Relating to the Garbage Industry of New Jersey* (Trenton, NJ: 1969), 4–5; cited in Musto and Commission, *Solid Waste: A Coordinated Approach*, 48.

13 Krumholz "Legal Issues in Solid Waste Management," 3.

14 Krumholz "Legal Issues in Solid Waste Management," 3.

15 Musto and Commission, *Solid Waste: A Coordinated Approach*, 48–49.

16 *Ringlieb v. Township of Parsippany-Troy Hills* 59 N.J. 348, 350 (1971), https://law.justia.com/cases/new-jersey/supreme-court/1971/59-n-j-348-0.html.

17 Oral opinion of Judge Joseph H. Stamler in *Ringlieb v. Township of Parsippany-Troy Hills* 59 N.J. 348, 350 (1971), https://law.justia.com/cases/new-jersey/supreme-court/1971/59-n-j-348-0.html.

18 These positions were later upheld in subsequent court cases, including *Little Falls Tp. v. Bardin, 173 N.J. Super. 397* (App.Div. 1979); as well as *Township of Chester v. Department of Environmental Protection of the State of New Jersey*, 181 N.J. Super. 445 (App. Div. 1981).

19 Musto and Commission, *Solid Waste: A Coordinated Approach*, x.

20 Musto and Commission, *Solid Waste: A Coordinated Approach*, vii.

21 Cited in Waggoner, "State Orders Tight Curbs on Outside Trash Haulers," 67.

22 Cited in Waggoner, "State Orders Tight Curbs on Outside Trash Haulers," 67.

23 HMDC regulatory code: NJAC 19:7-1.1(g) and (h), since amended.

24 Ch. 39 of NJ Laws of 1973, 95, http://njlaw.rutgers.edu/cgi-bin/diglib.cgi?collect=njleg&file=195_2&page=0095&zoom=150.

25 *Hackensack Meadowlands Devel. Comm'n v. Mun. Landfill Auth.* 127 N.J. Super. 160 (1974), https://law.justia.com/cases/new-jersey/appellate-division-published/1974/127-n-j-super-160-0.html.

26 *Hackensack Meadowlands Devel. Comm'n v. Mun. Landfill Auth.* (1974).

27 *Hackensack Meadowlands Devel. Comm'n v. Mun. Landfill Auth.* 68 N.J. 451 (1975), https://law.justia.com/cases/new-jersey/supreme-court/1975/68-n-j-451-0.html. The city of Philadelphia along with landfill operators in southern New Jersey had separately filed suit against the state of New Jersey and the NJDEP, with a similar outcome. The 1975 *Hackensack Meadowlands Devel. Comm'n v. Mun. Landfill Auth.* (68 N.J. 451) was a combined case heard before the Supreme Court of New Jersey, since the issues at stake were essentially identical.

28 *Hackensack Meadowlands Devel. Comm'n v. Mun. Landfill Auth.* 68 N.J. 451 (1975).

29 *Hackensack Meadowlands Devel. Comm'n v. Mun. Landfill Auth.* 68 N.J. 451 (1975).

30 *United States v. Pennsylvania Refuse Removal Ass'n.*, 242 F. Supp. 794 (E.D. Pa. 1965), aff'd, 357 F.2d 806 (3d Cir.), cert. denied, 384 U.S. 961, 86 S. Ct. 1588, 16 L. Ed. 2d 674 (1966). Cited in *Hackensack Meadowlands Devel. Comm'n v. Mun. Landfill Auth.* 68 N.J. 451 (1975).

31 *City of Philadelphia v. New Jersey* (No. 77–404) 437 U.S. 617 (1978), https://
www.law.cornell.edu/supremecourt/text/437/617#writing-USSC_CR_0437
_0617_ZD.

32 *City of Philadelphia v. New Jersey* (No. 77–404) 437 U.S. 617 (1978).

33 *City of Philadelphia v. New Jersey* (No. 77–404) 437 U.S. 617 (1978).

34 Musto and Commission, *Solid Waste: A Coordinated Approach*, 42.

35 Musto and Commission, *Solid Waste: A Coordinated Approach*, 29.

36 Musto and Commission, *Solid Waste: A Coordinated Approach*, 30.

37 and Commission 30.

38 Ch. 242 of NJ Laws of 1970, 849 et seq, http://njlaw.rutgers.edu/cgi-bin/diglib.cgi
?collect=njleg&file=194_1&page=0851&zoom=150.

39 Musto and Commission, *Solid Waste: A Coordinated Approach*, 50.

40 Musto and Commission, *Solid Waste: A Coordinated Approach*, 30.

41 Musto and Commission, *Solid Waste: A Coordinated Approach*, 42.

42 Ch. 326 of NJ Laws of 1975, 1279, http://njlaw.rutgers.edu/cgi-bin/diglib.cgi
?collect=njleg&file=196_2&page=1279&zoom=120.

43 Ch. 326 of NJ Laws of 1975, 1291–1295.

44 Ch. 326 of NJ Laws of 1975, 1280.

45 Musto and Commission, *Solid Waste: A Coordinated Approach*, 44.

46 Ch. 326 of NJ Laws of 1975, 1279.

47 Ch. 326 of NJ Laws of 1975, 1287.

48 Ch. 326 of NJ Laws of 1975, 1288.

49 Ch. 326 of NJ Laws of 1975, 1290.

50 Ch. 326 of NJ Laws of 1975, 1296.

51 Officially, the 326 amendments would be known as "An act concerning solid
waste management and resource recovery; designating solid waste management
districts within the State and regulating solid waste collection and disposal
therein; creating an Advisory Council on Solid Waste Management in the State
Department of Environmental Protection, and relating to the department's
functions, power and duties." Ch. 326 of NJ Laws of 1975, 1278.

52 See Burlington County Board of Chosen Freeholders, *The Plan for Solid Waste
Management in Burlington County, New Jersey* (Moorestown, NJ: Applied
Information Industries, 1972) and subsequent updates and amendments.

53 Division of Waste Management, *Solid Waste Management Plan*, A12–A13.

54 Division of Waste Management, *Solid Waste Management Plan*, 144.

55 See the Camden County summary, Division of Waste Management, *Solid Waste
Management Plan*, A8.

56 Division of Waste Management, *Solid Waste Management Plan*, A23–A24.

57 See summaries of Warren, Hunterdon, Sussex, and Morris County plans in
Division of Waste Management, *Solid Waste Management Plan*.

58 Division of Waste Management, *Solid Waste Management Plan*, A47–A48.

59 Division of Waste Management, *Solid Waste Management Plan*, 80.

60 These landfills were a mixture of publicly and privately owned: Hackensack
Meadowlands Development Commission 1-C; Hackensack Meadowlands
Development Commission Balefill; Kinsley Landfill, Inc. (Gloucester County);
Kingsland Park Sanitation Extension (Bergen County); Edgeboro Disposal, Inc.
(Middlesex County); Monmouth County Reclamation Center; Parklands
Reclamation (Burlington County); Pennsauken Sanitary Landfill (Camden
County); Pinelands Park Landfill (Atlantic County); Landfill and Development

Co. (Burlington County); and Ocean County Landfill. See Division of Waste Management, *Solid Waste Management Plan*, 71. See also New Jersey Institute of Technology and New Jersey Alliance for Action, *Solving the Garbage Disposal Crisis in New Jersey: A Cooperative Study* (Newark: New Jersey Institute of Technology and the New Jersey Alliance for Action, 1986), 3–1.

61 Musto and Commission, *Solid Waste: A Coordinated Approach*, 48–49.

62 Krumholz "Legal Issues in Solid Waste Management," 4.

63 John J. Degnan and Nathan M. Edelstein, Formal Opinion No. 3–1980 (Trenton: Office of the Attorney General of the State of New Jersey, 1980a), 1, http://njlegallib.rutgers.edu/legallib/ag/fo1980/no03.pdf.

64 Degnan and Edelstein, Formal Opinion No. 3–1980, 2.

65 Degnan and Edelstein, Formal Opinion No. 3–1980, 3–4.

66 John J. Degnan and Nathan M. Edelstein, Formal Opinion No. 12–1980 (Trenton: Office of the Attorney General of the State of New Jersey, 1980b, 3, http://njlegallib.rutgers.edu/legallib/ag/fo1980/no12.pdf.

67 Degnan and Edelstein, Formal Opinion No. 12–1980, 3.

68 *A.A. Mastrangelo, Inc. v. Environmental Protec. Dep't* (Short Title) 90 N.J. 666 (1982), https://casetext.com/case/aa-mastrangelo-inc-v-environmental-protec-dept.

69 From opinion of Judge Robert L. Clifford in *A.A. Mastrangelo, Inc. v. Environmental Protec. Dep't* (Short Title) 90 N.J. 666 (1982).

70 From opinion of Judge Robert L. Clifford in *A.A. Mastrangelo, Inc. v. Environmental Protec. Dep't* (Short Title) 90 N.J. 666 (1982).

71 Krumholz "Legal Issues in Solid Waste Management," 5–6.

72 Division of Waste Management, *Solid Waste Management Plan*, 69.

73 Division of Waste Management, *Solid Waste Management Plan*, 68–69.

74 See Assembly Agriculture and Environment Committee, *Public Hearing before Assembly Agriculture and Environment Committee on "Solid Waste Management & Interdistrict Waste Flow Orders"* (Trenton, NJ: Office of Legislative Services, 1985).

75 Division of Waste Management, *Solid Waste Management Plan*, A18–A19. Regarding the new, de facto ban on disposal of out-of-state wastes, by 1985 the "tipping fee" to use New Jersey disposal facilities had grown so large that the majority of New York and Pennsylvania waste haulers were no longer utilizing them anyway; see Division of Waste Management, *Solid Waste Management Plan*, 71.

76 John J. Bergin, and Office of the Deputy Attorney General, *Garbage Collection Practices* (Trenton, NJ: Office of the Attorney General, 1959), 17.

77 *United States v. Pennsylvania Refuse Removal Ass'n, Harry Coren, Arnold Graf, Salvatore Graziano, and Edwin S. Vile* 242 F. Supp. 794 (E.D. Pa., 1965), https://casetext.com/case/united-states-v-pennsylvania-refuse-removal-assn.

78 From opinion of Judge Alfred L. Luongo in *United States v. Pennsylvania Refuse Removal Ass'n* (1965).

79 From opinion of Judge Alfred L. Luongo in *United States v. Pennsylvania Refuse Removal Ass'n* (1965).

80 All of these cases are cited and discussed in Task Force to Study Solid Waste Regulation, *Report of the Departments of Law and Public Safety and Environmental Protection and the Board of Public Utilities Task Force to Study Solid Waste Regulation* (Trenton: New Jersey Board of Public Utilities and Departments of Environmental Protection and Law and Public Safety, 1981), 5.

81 Task Force to Study Solid Waste Regulation, *Report of the Departments of Law and Public Safety*, 4.

82 Task Force to Study Solid Waste Regulation, *Report of the Departments of Law and Public Safety*, 19–22.

83 The hearings and investigations become quite repetitive in content; for the culminating hearing which considers most of the major issues in this area see Solid Waste Subcommittee of the Assembly Agriculture and Environment Committee, *Public Hearing before Solid Waste Subcommittee of the Assembly Agriculture and Environment Committee on Illegal and Unethical Practices in the Solid Waste Industry: held September 15 and October 1 1981, Trenton, New Jersey* (Trenton: New Jersey Legislature, 1981).

84 No author given, "2 Jersey Garbage Haulers Convicted of Conspiracy," *New York Times*, April 10, 1983, 001040.

85 Interested readers should see New Jersey State Commission of Investigation, *Solid Waste Regulation* (Trenton: New Jersey State Commission of Investigation, 1989).

86 This book focuses on the policy surrounding the waste management industry in New Jersey and therefore I do not wish to consider the individual episodes of corruption and violence that unfolded during the 1970s and 1980s in much detail. At the same time, I do not wish to minimize the seriousness of these crimes, and it should be clear to all readers that such approaches to 'doing business' have no place in a civil society.

87 New Jersey State Commission of Investigation, *Solid Waste Regulation*, 9–10.

88 New Jersey State Commission of Investigation, *Solid Waste Regulation*, 47.

89 New Jersey State Commission of Investigation, *Solid Waste Regulation*, 47.

90 New Jersey State Commission of Investigation, *Industrious Subversion: Circumvention of Oversight in Solid Waste and Recycling in New Jersey* (Trenton, NJ: New Jersey State Commission of Investigation, 2011), 12–13.

91 See New Jersey State Commission of Investigation, *Industrious Subversion*; see also Tom Johnson and John Mooney, "Dirty Business: Report Blasts Mob Involvement in Garbage and Recycling Operations," *NJ Spotlight*, December 7, 2011, https://www.njspotlight.com/stories/11/1206/2356/.

92 New Jersey State Commission of Investigation, *Industrious Subversion*, 29–31.

93 See New Jersey State Commission of Investigation, *Solid Waste Regulation*; New Jersey State Commission of Investigation, *Solid Waste Management by the Bergen County Utilities Authority* (Trenton: New Jersey State Commission on Investigation, 1992); New Jersey Department of Law and Public Safety, *Annual Report to Governor on Environmental Crime Prosecutions* (Trenton: New Jersey Department of Law and Public Safety, 2006); New Jersey State Commission of Investigation, *Industrious Subversion*; New Jersey State Commission of Investigation, *Dirty Dirt: The Corrupt Recycling of Contaminated Soil and Debris* (Trenton: New Jersey State Commission of Investigation, 2016); and New Jersey State Commission of Investigation, *Dirty Dirt II: Bogus Recycling of Tainted Soil and Debris* (Trenton: New Jersey State Commission of Investigation, 2019).

94 New Jersey State Commission of Investigation, *Dirty Dirt*, 4.

95 See New Jersey State Commission of Investigation, *Dirty Dirt*; New Jersey State Commission of Investigation, *Dirty Dirt II*; Michael Sol Warren, "Mob-Tied Illegal Dumping in N.J. May Finally End After Years of Warning," *NJ.com*, June 26, 2019, https://www.nj.com/news/2019/06/mob-tied-illegal-dumping-in-nj-may-finally-end-after-years-of-warning.html.

96 New Jersey State Commission of Investigation, *Solid Waste Regulation*, 4.

97 See New Jersey State Commission of Investigation, *Solid Waste Regulation*; also New Jersey Department of Law and Public Safety, *Annual Report to Governor on Environmental Crime Prosecutions*.

98 New Jersey State Commission of Investigation, *Solid Waste Regulation*, 38–39.

99 Cited in New Jersey State Commission of Investigation, *Solid Waste Regulation*, 39.

100 Division of Waste Management, *Solid Waste Management Plan*, 86.

101 Bergin, and Office of the Deputy Attorney General, *Garbage Collection Practices*, 15.

Chapter 4   Recycle or Incinerate?

Epigraph Source: Thomas H. Kean, "Annual Message to the New Jersey State Legislature" (Address to the New Jersey State Legislature, Trenton, NJ, 1985), 28–29.

1 Kean, "Annual Message to the New Jersey State Legislature," 15–16.

2 William V. Musto, *Solid Waste: A Coordinated Approach* (Trenton, NJ: County and Municipal Government Study Commission, 1972), ix–x.

3 Musto, *Solid Waste*, ix.

4 Officially, the 326 amendments would be known as "An act concerning solid waste management and resource recovery; designating solid waste management districts within the State and regulating solid waste collection and disposal therein; creating an Advisory Council on Solid Waste Management in the State Department of Environmental Protection, and relating to the department's functions, power and duties." Ch. 326 of NJ Laws of 1975, 1278.

5 Ch. 326 of NJ Laws of 1975, 1278

6 Brendan Byrne, "The Fifth Annual Message Delivered to the Legislature" (Address to the New Jersey State Legislature, Trenton, NJ, 1979), 6.

7 Byrne, "The Fifth Annual Message Delivered to the Legislature," 10.

8 Division of Waste Management, NJ Department of Environmental Protection, "Progress in Waste Management: A Solution to New Jersey's Garbage Dilemma" (Trenton: New Jersey Department of Environmental Protection, 1984).

9 Thomas H. Kean, "Annual Message to the New Jersey State Legislature" (Address to the New Jersey State Legislature, Trenton, NJ, 1988), 55–56.

10 Democratic State Committee, "1989 Democratic Platform" (Piscataway, NJ: 1989), 3–4.

11 James J. Florio, *Jim Florio on Solid Waste Disposal* (Mt. Laurel, NJ: Florio for Governor, Inc., 1989). Emphasis in original.

12 Florio, *Jim Florio on Solid Waste Disposal*, 6.

13 Florio, *Jim Florio on Solid Waste Disposal*, 6.

14 See Kat Eschner, "How the 1970s Created Recycling As We Know It," *Smithsonian Magazine*, November 15, 2017, https://www.smithsonianmag.com/smart-news/how-1970s-created-recycling-we-know-it-180967179/; also Sarah Goodyear, "A Brief History of Household Recycling," *Citylab*, 2015, https://www.citylab.com/city-makers-connections/recycling/#slide-opener, accessed December 16, 2019.

15 Wayne DeFeo, interview by Jordan P. Howell, November 28, 2017.

16 See the Division of Waste Management, NJ Department of Environmental Protection, *Solid Waste Management Plan: Draft Update, 1985–2000* (Trenton, NJ: New Jersey Department of Environmental Protection, 1985), 42–43; also Robert Cowles Letcher and Mary T. Sheil, "Source Separation and Citizen Recycling," in

The Solid Waste Handbook: A Practical Guide, ed. William D. Robinson (Hoboken, NJ: John Wiley & Sons, Inc., 1986).

17 Letcher and Sheil, "Source Separation and Citizen Recycling," 246.

18 New Jersey Advisory Committee on Recycling, *Recycling in the 1980's: The Report on Recycling in New Jersey*. (Newark: Office of Recycling, New Jersey Department of Environmental Protection and Department of Energy, 1980).

19 Letcher and Sheil, "Source Separation and Citizen Recycling," 244.

20 Ch. 278 of NJ Laws of 1981, 1000 et seq, http://njlaw.rutgers.edu/cgi-bin/diglib.cgi ?collect=njleg&file=199_2&page=1000&zoom=120, accessed December 17, 2019.

21 Ch. 278 of NJ Laws of 1981, 1000.

22 Ch. 278 of NJ Laws of 1981.

23 Letcher and Sheil, "Source Separation and Citizen Recycling," 240.

24 Division of Waste Management, *Solid Waste Management Plan: Draft Update*, 43.

25 Letcher and Sheil, "Source Separation and Citizen Recycling," 246.

26 Letcher and Sheil, "Source Separation and Citizen Recycling," 246–247; see also Office of Recycling, Division of Waste Management, *Statewide Survey of Recycling Programs* (Trenton: New Jersey Department of Environmental Protection, 1982).

27 See, for instance, New Jersey Division of Environmental Quality, *Recycling: An Instructional and Informational guide* (Trenton: New Jersey Department of Environmental Protection, 1972); New Jersey Solid Waste Advisory Council, *Encouraging Recycling in New Jersey: A First Step* (Trenton: New Jersey Solid Waste Advisory Council, 1973); New Jersey Advisory Council on Solid Waste Management, *Report by the Council on its Public Hearing on Source Separation and Recycling* (Trenton: New Jersey Advisory Council on Solid Waste Management, 1978); New Jersey Department of Energy and New Jersey Department of Environmental Protection, *New Jersey Recyclers' Guide* (Trenton: New Jersey Department of Energy and New Jersey Department of Environmental Protection, 1979).

28 There are too many examples to list in their entirety, but see the following as examples of this publication and education drive: Office of Recycling, Division of Waste Management, *Municipal Budgets and Recycling* (Newark: New Jersey Department of Environmental Protection, 1982); Office of Recycling, Division of Waste Management, *Used Oil, It's Easy to Recycle: Today's Conservation is Tomorrow's Resource* (Newark: New Jersey Department of Environmental Protection, 1982); Office of Recycling, Division of Waste Management, *Recycling in the 1980's: Progress Report and Program Recommendations* (Newark: New Jersey Department of Environmental Protection, 1984); Office of Recycling, Division of Waste Management, *Getting the Word Out: A Guide to Publicity* (Trenton: New Jersey Department of Environmental Protection, 1986); Mary T. Sheil, *Steps in Organizing a Municipal Recycling Program* (Trenton: New Jersey Department of Environmental Protection, 1986); Peter F. Strom and Melvin S. Finstein, *Leaf Composting Manual for New Jersey Municipalities* (Trenton: New Jersey Department of Environmental Protection, 1986); Division of Solid Waste Management, New Jersey Department of Environmental Protection, *What Would You Do with 11,000,000 Tons of Garbage? Solid Waste Management* (Trenton: New Jersey Department of Environmental Protection, 1989).

29 For example, Office of Recycling, Division of Waste Management, *Recycling Roundups* (Trenton: New Jersey Department of Energy and New Jersey Department of Environmental Protection, 1981); also Office of Recycling, Division of

Waste Management, *Recycle-gram* (Newark: New Jersey Department of Environmental Protection, 1984).

30 New Jersey Division of Environmental Quality, *Recycling: An Instructional and Informational guide* (Trenton: New Jersey Department of Environmental Protection, 1972), 3.

31 For example, Office of Recycling, Division of Waste Management, *Getting the Word Out: A Guide to Publicity* (Trenton: New Jersey Department of Environmental Protection, 1986); also Office of Recycling, Division of Waste Management, *A Guide to Marketing Recyclable Materials* (Trenton: New Jersey Department of Environmental Protection, 1987).

32 Arthur D. Little Inc., *Marketing Development Strategies for Recyclable Materials: Batteries, Waste Paper, Plastics, Ferrous Auto Scrap, Tires* (Trenton: Office of Recycling; New Jersey Department of Environmental Protection, 1988), 1–1.

33 See Assembly Energy and Natural Resources Committee, *Public Hearing before Assembly Energy and Natural Resources Committee on Beverage Container Deposit Legislation* (Trenton: New Jersey Legislature, 1982); New Jersey General Assembly, *General Assembly Forum on the Beverage Container Deposit and Refund Act and Related Issues: A-(1753/78/182/316/482/533/1180/1237/1671) ACS (Paterniti), Held December 1, 1983, Trenton, New Jersey* (Trenton: New Jersey Legislature, 1983); Assembly Agriculture and Environment Committee, *Public Hearing before Assembly Agriculture and Environment Committee on Assembly bill 2606: (Requires a 10% Refund Value on Certain Beverage Containers at the Point of Distribution): Held February 5, 1984, Trenton, New Jersey* (Trenton: New Jersey Legislature, 1985a); and Assembly Agriculture and Environment Committee, *Public Hearing before Assembly Agriculture and Environment Committee on Assembly Bill 3382: (Establishes a Mandatory Statewide Recycling Program) and Assembly Bill 3398 (Designated the "Recycling and Beverage Container Redemption Act"): Held May 2, 1985, Trenton, New Jersey* (Trenton: New Jersey Legislature, 1985b).

34 For an interesting study of bottle deposit schemes and the development of "reverse vending machines," see Finn Arne Jørgensen, *Making a Green Machine: The Infrastructure of Beverage Container Recycling* (New Brunswick, NJ: Rutgers University Press. 2011).

35 New Jersey General Assembly, *General Assembly Forum on the Beverage Container Deposit*, 1–2.

36 New Jersey General Assembly, *General Assembly Forum on the Beverage Container Deposit*, 4–5.

37 New Jersey General Assembly, *General Assembly Forum on the Beverage Container Deposit*, 5–6.

38 New Jersey General Assembly, *General Assembly Forum on the Beverage Container Deposit*, 21.

39 New Jersey General Assembly, *General Assembly Forum on the Beverage Container Deposit*, 56.

40 New Jersey General Assembly, *General Assembly Forum on the Beverage Container Deposit*, 56.

41 New Jersey General Assembly, *General Assembly Forum on the Beverage Container Deposit*, 7.

42 New Jersey General Assembly, *General Assembly Forum on the Beverage Container Deposit*, 8.

43 New Jersey General Assembly, *General Assembly Forum on the Beverage Container Deposit*, 8.

44 New Jersey General Assembly, *General Assembly Forum on the Beverage Container Deposit*, 8–9.

45 New Jersey General Assembly, *General Assembly Forum on the Beverage Container Deposit*, 46.

46 New Jersey General Assembly, *General Assembly Forum on the Beverage Container Deposit*, 47.

47 New Jersey General Assembly, *General Assembly Forum on the Beverage Container Deposit*, 47.

48 New Jersey General Assembly, *General Assembly Forum on the Beverage Container Deposit*, 94.

49 New Jersey General Assembly, *General Assembly Forum on the Beverage Container Deposit*, 95.

50 New Jersey General Assembly, *General Assembly Forum on the Beverage Container Deposit*, 23.

51 New Jersey General Assembly, *General Assembly Forum on the Beverage Container Deposit*, 23.

52 New Jersey General Assembly, *General Assembly Forum on the Beverage Container Deposit*, 28–29.

53 Assembly Agriculture and Environment Committee, *Public Hearing before Assembly Agriculture and Environment Committee on Assembly bill 2606*, 26.

54 Assembly Agriculture and Environment Committee, *Public Hearing before Assembly Agriculture and Environment Committee on Assembly bill 2606*, 27.

55 Assembly Agriculture and Environment Committee, *Public Hearing before Assembly Agriculture and Environment Committee on Assembly bill 2606*, 28.

56 Assembly Agriculture and Environment Committee, *Public Hearing before Assembly Agriculture and Environment Committee on Assembly bill 2606*, 31–32.

57 Assembly Agriculture and Environment Committee, *Public Hearing before Assembly Agriculture and Environment Committee on Assembly bill 2606*, 37–39.

58 Assembly Agriculture and Environment Committee, *Public Hearing before Assembly Agriculture and Environment Committee on Assembly bill 2606*, 37–40.

59 Arthur D. Little Inc., *Marketing Development Strategies*, 1–14.

60 New Jersey General Assembly, *General Assembly Forum on the Beverage Container Deposit*, 45–46.

61 Division of Waste Management, *Solid Waste Management Plan: Draft Update*, appendix C.

62 Assembly Agriculture and Environment Committee, *Public Hearing before Assembly Agriculture and Environment Committee on Assembly Bill 3382*, 3.

63 Assembly Agriculture and Environment Committee, *Public Hearing before Assembly Agriculture and Environment Committee on Assembly Bill 3382*, page 4 of included draft bill A-3382.

64 Senate Energy and Environment Committee, *Public Hearing before Senate Energy and Environment Committee on Senate Bill 2820: (Expands and Extends Existing "Recycling Act")* (Trenton: Office of Legislative Services, 1985) page 4 of included draft bill S-2820.

65 Assembly Agriculture and Environment Committee, *Public Hearing before Assembly Agriculture and Environment Committee on Assembly Bill 3382*, 12.

66 Assembly Agriculture and Environment Committee, *Public Hearing before Assembly Agriculture and Environment Committee on Assembly Bill 3382*, 10–11.

67 Assembly Agriculture and Environment Committee, *Public Hearing before Assembly Agriculture and Environment Committee on Assembly Bill 3382*; see also Senate Energy and Environment Committee, *Public Hearing before Senate Energy and Environment Committee on Senate Bill 2820*; and Senate Energy and Environment Committee, *Public Hearing before the Senate Energy and Environment Committee on S-1478 (Statewide Mandatory Recycling)* (Trenton: Office of Legislative Services, 1986).

68 Assembly Agriculture and Environment Committee, *Public Hearing before Assembly Agriculture and Environment Committee on Assembly Bill 3382*, 57.

69 Assembly Agriculture and Environment Committee, *Public Hearing before Assembly Agriculture and Environment Committee on Assembly Bill 3382*, 58.

70 Assembly Agriculture and Environment Committee, *Public Hearing before Assembly Agriculture and Environment Committee on Assembly Bill 3382*, 50–53.

71 Ch. 102 of NJ Laws of 1987, 359 et seq, http://njlaw.rutgers.edu/cgi-bin/diglib.cgi ?collect=njleg&file=202_2&page=0359&zoom=70, accessed January 4, 2020.

72 Ch. 102 of NJ Laws of 1987, 360.

73 Arthur D. Little Inc., *Marketing Development Strategies*, 1–3.

74 Letcher and Sheil, "Source Separation and Citizen Recycling," 247.

75 Assembly Agriculture and Environment Committee, *Public Hearing before Assembly Agriculture and Environment Committee on Assembly Bill 3382*, 52.

76 Letcher and Sheil, "Source Separation and Citizen Recycling," 247.

77 Letcher and Sheil, "Source Separation and Citizen Recycling," 247.

78 See Clean Air Council, *There's More Than One Way to Recycle: Case Studies of Recycling Programs in Camden County, New Jersey, Chicago, Illinois, Kitchener, Ontario, Canada, Marin County, California, Minneapolis, Minnesota, Montclair, New Jersey, Spring City, Pennsylvania, Toledo, Ohio* (Philadelphia, PA: Clean Air Council, 1989), 1–4.

79 Letcher and Sheil, "Source Separation and Citizen Recycling," 247.

80 Office of Recycling, Division of Waste Management, *Managing the Solid Waste Crisis in New Jersey: New Jersey Statewide Mandatory Source Separation and Recycling Act Summary*, P.L. 1987, c.102 revised September 2, 1987 (Trenton: New Jersey Department of Environmental Protection, 1987), 5–6.

81 New Jersey State Commission of Investigation, *Solid Waste Regulation* (Trenton: New Jersey State Commission of Investigation, 1989), 41.

82 New Jersey State Commission of Investigation, *Industrious Subversion: Circumvention of Oversight in Solid Waste and Recycling in New Jersey* (Trenton, NJ: New Jersey State Commission of Investigation, 2011), 38.

83 New Jersey State Commission of Investigation, *Industrious Subversion*, 38.

84 New Jersey State Commission of Investigation, *Industrious Subversion*, 1–2.

85 New Jersey State Commission of Investigation, *Industrious Subversion*, 39.

86 Assembly Agriculture and Environment Committee, *Public Hearing before Assembly Agriculture and Environment Committee on Assembly Bill 3382*, 1–2 of testimony.

87 Musto, *Solid Waste*, ix-x.

88 Quotations from original sources are lightly edited to reflect these various phrasings, most commonly through the use of bracketed text, e.g., [WTE] so that

the distinctions between different technologies are clear to readers. Consult original sources for original phrasings.

89 Commission to Study the Problem of Solid Waste Disposal, *Public Hearing before Commission to Study the Problem of Waste Disposal: (Created under Provisions of Assembly Concurrent Resolution no. 36)* (Trenton: New Jersey Legislature, 1965), 4.

90 Musto, *Solid Waste*, 21.

91 Commission to Study the Problem of Solid Waste Disposal, *Public Hearing*, 4.

92 Planners Associates Inc., *New Jersey State Solid Waste Management Plan* (Newark: Planners Associates Inc. and Bureau of Solid Waste Management, New Jersey Department of Environmental Protection, 1970), 96.

93 This is a simplified version of the history of incineration and WTE; in the late nineteenth and early twentieth centuries, British engineers also attempted to capture some of the waste heat from incineration, but with mixed results. For a more complete accounting of incineration and WTE, see Jordan P. Howell, "Technology and Place: A Geography of Waste-to-Energy in the United States" (PhD diss., Michigan State University, 2013); Jordan P. Howell, "Considering Waste-to-Energy Facilities in the US" (working paper, Rowan Solid Waste Lab, Rowan University, 2014), http://www.rowan.edu/colleges/chss/departments /geography/rswl/resources/whitepapers.html; Jordan P. Howell, "Sweetness and HPOWER: Waste, Sugar, and Ecological Identity in the Development of Honolulu's HPOWER Waste-to-Energy Facility," *Global Environment* 13 (2020): 285–316; Martin V. Melosi, "Technology Diffusion and Refuse Disposal: The Case of the British Destructor," in *Technology and the Rise of the Networked City in Europe and America*, ed. Joel Tarr and Gabriel Dupuy (Philadelphia, PA: Temple University Press, 1988), 207–226; and Martin V. Melosi, *Garbage in the Cities: Refuse, Reform, and the Environment*, rev. ed (Pittsburgh, PA: University of Pittsburgh Press, 2005).

94 New Jersey Department of Energy, *Energy Master Plan Preliminary Policy Statement: Solid Waste, Its Energy Conservation and Production Potential* (Trenton: New Jersey Department of Energy, 1978).

95 New Jersey Department of Energy, *Energy Master Plan*, 1.

96 New Jersey Department of Energy, *Energy Master Plan*, 6.

97 New Jersey Department of Energy, *Energy Master Plan*, 12–13.

98 Essex County Division of Solid Waste Management, *The Integration of Energy & Material Recovery in the Essex County Solid Waste Management Program* (Belleville, NJ: Essex County Division of Solid Waste Management, 1983), 1.

99 Essex County Division of Solid Waste Management, *The Integration of Energy & Material Recovery*, 1.

100 Division of Waste Management, *Progress in Waste Management: A Solution to New Jersey's Garbage Dilemma* (Trenton: New Jersey Department of Environmental Protection, 1984).

101 Division of Waste Management, *Progress in Waste Management*, 2.

102 Division of Waste Management, *Solid Waste Management Plan: Draft Update, 1985–2000* (Trenton: New Jersey Department of Environmental Protection, 1985), 12.

103 Division of Waste Management, *Solid Waste Management Plan*, 2–10.

104 Assembly County Government and Regional Authorities Committee, *Public Hearing before Assembly County Government and Regional Authorities*

Committee on Assembly bill 1778: (Provides for a Resource Recovery Investment Tax on Solid Waste Disposal at Sanitary Landfills) (Trenton, NJ: New Jersey Legislature, 1984), 2–3.

105 Dan Kirshner and Adam C. Stern, *To Burn or Not to Burn: The Economic Advantages of Recycling over Garbage Incineration for New York City* (New York, Environmental Defense Fund, 1985).

106 Kirshner and Stern, *To Burn or Not to Burn*, vi.

107 Kirshner and Stern, *To Burn or Not to Burn*, 3.

108 As just a sampling of these planning documents and studies, see: New Jersey Advisory Council on Solid Waste Management, *Public Hearing Report: "What Must Be Accomplished to Implement Energy Recovery Facilities as Approved in the New Jersey Solid Waste Management District Plans"* (Trenton: New Jersey Advisory Council on Solid Waste Management, 1983); *State of New Jersey Incinerator Study*, ed. Robert G. McInnes and Robert R. Hall, vol. 3, *Technical Review and Regulatory Analysis of Municipal Incineration* (Washington, DC: GCA Corporation, GCA/Technology Division, 1984); New Jersey Clean Air Council, New Jersey Advisory Council on Solid Waste Management, *Report and Recommendations Based on a Joint Public Hearing Held March 19, 1984 by the New Jersey Clean Air Council and the New Jersey Advisory Council on Solid Waste Management on "The Effect of Resource Recovery Technologies on Air Quality"* (Trenton, NJ: New Jersey Department of Environmental Protection, 1984); Susan C. Remis, "White paper of the Department of the Public Advocate, Division of Public Interest Advocacy on the Selection of a Resource Recovery Plant Site: A Case Study of Mercer County's Site Selection Process" (Trenton: New Jersey Division of Public Interest Advocacy, 1984); Marc Jay Rogoff, *How to Implement Waste-to-Energy Projects* (Park Ridge, NJ: Noyes Publications, 1987); Assembly Solid Waste Management Committee, *Public Hearing before Assembly Solid Waste Management Committee: Assembly Bill no. 3107 (2R), (Authorizes $135 Million in General Obligation Bonds for Construction of Resource Recovery Facilities and Environmentally Sound Sanitary Landfill Facilities)* (Trenton, NJ: Office of Legislative Services, 1988); Assembly County Government and Regional Authorities Committee, *Public Meeting before Assembly County Government and Regional Authorities Committee: Assembly Bill no. 4105 (Requires DEP Study of Cumulative Impact of Resource Recovery Facilities on the Environment, Appropriates $75,000)*, vol. 1. (Trenton, NJ: Office of Legislative Services, 1989); and Senate Energy and Environment Committee, *Public Hearing before Senate Energy and Environment Committee: To Discuss the Appropriate Role of Incineration, and its Alternatives, in the State's Long-Term Disposal Strategy* (Trenton, NJ: Office of Legislative Services, 1989).

109 Assembly Select Committee on Solid Waste Disposal, *Public Meeting before Assembly Select Committee on Solid Waste Disposal: Assembly Bill 3892 (Authorizes DEP to Implement Emergency Plans on Behalf of Counties during a Declared State of Solid Waste Emergency)* (Trenton: New Jersey Legislature, 1987).

110 Assembly Select Committee on Solid Waste Disposal, *Public Meeting*, 2–3.

111 Assembly Select Committee on Solid Waste Disposal, *Public Meeting*, appendix, ix–x. Emphasis in original.

112 Republican Party (NJ), "1989 Republican Platform" (Trenton, NJ: 1989).

Chapter 5   Limits to the System

Epigraph Source: Written testimony of Albert A. Fiore, offered to Senate Energy and Environment Committee, *Public Hearing before Senate Energy and Environment Committee: To Discuss the Appropriate Role of Incineration, and its Alternatives, in the State's Long Term Disposal Strategy* (Trenton, NJ: Office of Legislative Services, 1989), 42x.

  1 See the reprint of Governor's Executive Order no. 8 of 1990, published in Judith A. Yaskin, and Emergency Solid Waste Assessment Task Force, *Final Report: Emergency Solid Waste Assessment Task Force* (Trenton: New Jersey Department of Environmental Protection, 1990).
  2 Senate Energy and Environment Committee, *Public Hearing before Senate Energy and Environment Committee: To Discuss the Appropriate Role of Incineration, and its Alternatives, in the State's Long Term Disposal Strategy* (Trenton, NJ: Office of Legislative Services, 1989), 1.
  3 Senate Energy and Environment Committee, *Public Hearing*, 10–11.
  4 Senate Energy and Environment Committee, *Public Hearing*, 13.
  5 Senate Energy and Environment Committee, *Public Hearing*, 19–20.
  6 Senate Energy and Environment Committee, *Public Hearing*, 27–31.
  7 Senate Energy and Environment Committee, *Public Hearing*, 44–45.
  8 Office of Recycling, Division of Waste Management, *Recycling Into the 90s: Recycling Report to the Governor and the Legislature as Required by N.J.S.A. 13:1E section, 50–99.11 et seq* (Trenton: New Jersey Department of Environmental Protection, 1990), i.
  9 Office of Recycling, *Recycling Into the 90s*.
 10 Office of Recycling, *Recycling Into the 90s*, 39.
 11 See Assembly Waste Management Planning and Recycling Committee, *Public Hearing before Assembly Waste Management, Planning and Recycling Committee: Recycling in New Jersey - Progress Report on the Recycling of Aluminum, Glass, Plastics and Newspapers* (Trenton, NJ: Office of Legislative Services, 1990), 6–7.
 12 Assembly Waste Management Planning and Recycling Committee, *Public Hearing*, 10.
 13 Assembly Waste Management Planning and Recycling Committee, *Public Hearing*, 14–17.
 14 Assembly Waste Management Planning and Recycling Committee, *Public Hearing*, 73–79.
 15 Yaskin and Emergency Solid Waste Assessment Task Force, *Final Report*, 2–3.
 16 Yaskin and Emergency Solid Waste Assessment Task Force, *Final Report*, 5.
 17 Yaskin and Emergency Solid Waste Assessment Task Force, *Final Report*, 6.
 18 Yaskin and Emergency Solid Waste Assessment Task Force, *Final Report*, 11–12.
 19 Yaskin and Emergency Solid Waste Assessment Task Force, *Final Report*, 11.
 20 Yaskin and Emergency Solid Waste Assessment Task Force, *Final Report*, 12.
 21 Yaskin and Emergency Solid Waste Assessment Task Force, *Final Report*, 12.
 22 Yaskin and Emergency Solid Waste Assessment Task Force, *Final Report*, 32.
 23 Yaskin and Emergency Solid Waste Assessment Task Force, *Final Report*, 34.
 24 Yaskin and Emergency Solid Waste Assessment Task Force, *Final Report*, 22–26.
 25 Division of Solid Waste Management, New Jersey Department of Environmental Protection and Energy, *Draft New Jersey State Solid Waste Management Plan Update 1993–2002, Section I: Municipal and Industrial Solid Waste* (Trenton, NJ:

New Jersey Department of Environmental Protection and Energy, 1993), executive summary, i.

26 Division of Solid Waste Management, *Draft New Jersey State Solid Waste Management Plan*, executive summary, 20.

27 Division of Solid Waste Management, *Draft New Jersey State Solid Waste Management Plan*, executive summary, 24.

28 Division of Solid Waste Management, *Draft New Jersey State Solid Waste Management Plan*, executive summary, 46–47.

29 Division of Solid Waste Management, *Draft New Jersey State Solid Waste Management Plan*, main section, 100–101.

30 Division of Solid Waste Management, *Draft New Jersey State Solid Waste Management Plan*, main section, 20.

31 See *CA Carbone, Inc. v. Clarkstown*, 511 U.S. 383 (1994), https://casetext.com/case /ca-carbone-inc-v-town-of-clarkstown-new-york.

32 See commentary prior to case text of *CA Carbone, Inc. v. Clarkstown*.

33 See "Opinion of the Court," *CA Carbone, Inc. v. Clarkstown*.

34 See *Atlantic Coast Demo. v. Board of Chosen Freeholders*, 112 F.3d 652 (3d Cir. 1997), https://casetext.com/case/atlantic-coast-demo-v-bd-chosen-freeholders.

35 See *Atlantic Coast Demolition v. Board of Chosen Freeholders*, 48 F.3d 701 (3d Cir. 1995), https://casetext.com/case/atlantic-coast-demolition-recycling-inc-v-board -of-chosen-freeholders-of-atlantic-county-4.

36 See *Atlantic Coast Demolition v. Board of Chosen Freeholders*, 48 F.3d 701 (3d Cir. 1995).

37 See *Waste Management of Pennsylvania, v. Shinn*, 938 F. Supp. 1243 (D.N.J. 1996), https://casetext.com/case/waste-management-of-pennsylvania-v-shinn.

38 See *Waste Management of Pennsylvania, v. Shinn*, 938 F. Supp. 1243 (D.N.J. 1996).

39 See *Waste Management of Pennsylvania, v. Shinn*, 938 F. Supp. 1243 (D.N.J. 1996).

40 See *Waste Management of Pennsylvania, v. Shinn*, 938 F. Supp. 1243 (D.N.J. 1996).

41 Gary Sondermeyer, interview by Jordan P. Howell, March 17, 2017.

42 John Turner, "The Flow Control of Solid Waste and the Commerce Clause: Carbone and Its Progeny," *Villanova Environmental Law Journal* 7, no. 2 (1996): 203–261, 261.

43 Senate Environment Committee, *September 19 Committee Meeting of Senate Environment Committee: Testimony Concerning the Effect of the United States District Court Decision in Atlantic Coast Demolition and Recycling, Inc., et al., v. Board of Chosen Freeholders of Atlantic County, et al., Declaring New Jersey's Waste Flow Regulations Unconstitutional* (Trenton: New Jersey Legislature, 1996), 1–2.

44 Division of Solid Waste Management, *Draft New Jersey State Solid Waste Management Plan*, executive summary, 49.

45 New Jersey Department of Environmental Protection, *Statewide Solid Waste Management Plan* (Trenton: New Jersey Department of Environmental Protection, 2006), F-1.

46 Division of Solid Waste Management, *Draft New Jersey State Solid Waste Management Plan*, executive summary, 49.

47 Senate Energy and Environment Committee, *Public Hearing*, 42x.

48 New Jersey State Commission of Investigation, *Solid Waste Regulation* (Trenton: New Jersey State Commission of Investigation, 1989), 38.

49 New Jersey State Commission of Investigation, *Solid Waste Regulation*, 40.

50 New Jersey State Commission of Investigation, *Solid Waste Regulation*, 38.

51 See Assembly Solid and Hazardous Waste Committee, *February 11th Public Hearing before Assembly Solid and Hazardous Waste Committee: Testimony from County Solid Waste Officials on the Evolution of County Solid Waste Management Activities Since the Waste Management v. Shinn, Carbone, and Atlantic Coast Decisions* (Trenton: New Jersey Legislature, 1999), 4–5.

52 See, for example, Assembly Agriculture and Waste Management Committee, *November 25th Committee Meeting of Assembly Agriculture and Waste Management Committee: Assembly bill no. 50 (Solid Waste Management and Environmental Investment Cost Recovery Act)* (Trenton: New Jersey Legislature, 1996); Assembly Agriculture and Waste Management Committee, *December 5th Committee Meeting of Agriculture and Waste Management Committee: Assembly bill no. 50 (Solid Waste Management and Environmental Investment Cost Recovery Act), Assembly Agriculture and Waste Management Committee* (Trenton: New Jersey Legislature, 1996); Senate Environment Committee, *September 19 Committee Meeting of Senate Environment Committee: Testimony Concerning the Effect of the United States District Court Decision in Atlantic Coast Demolition and Recycling, Inc., et al., v. Board of Chosen Freeholders of Atlantic County, et al., Declaring New Jersey's Waste Flow Regulations Unconstitutional* (Trenton: New Jersey Legislature, 1996); Senate Environment Committee, *October 28th Committee Meeting of Senate Environment Committee: The Effect of the United States District Court Decision Declaring New Jersey's Waste Flow Regulations Unconstitutional* (Trenton: New Jersey Legislature, 1996); Assembly Agriculture and Waste Management Committee, *January 23rd Committee Meeting of Agriculture and Waste Management Committee: Assembly Bill no. 50, Solid Waste Management and Environmental Investment Cost Recovery Act* (Trenton: New Jersey Legislature, 1997); Assembly Agriculture and Waste Management Committee, *March 25th Committee Meeting of Assembly Agriculture and Waste Management Committee: Assembly Bill nos. 50, 2568, 2811, and 2837* (Trenton: New Jersey Legislature, 1997); Assembly Agriculture and Waste Management Committee, *May 1st Committee Meeting of Assembly Agriculture and Waste Management Committee: Assembly Bill nos. 50 (ACS), 2568, 2811, and 2837 (Discussion on Bills Relating to Solid Waste Management and Disposal)* (Trenton: New Jersey Legislature, 1997); Assembly Agriculture and Waste Management Committee, *May 12th Committee Meeting of Assembly Agriculture and Waste Management Committee: Assembly Bill no. 50 (ACS), Solid Waste Management and Environmental Investment Cost Recovery Act* (Trenton: New Jersey Legislature, 1997); Assembly Agriculture and Waste Management Committee, *June 12th Committee Meeting of Assembly Agriculture and Waste Management Committee: Assembly Bill nos. 50 (Proposed ACS), 1171 (Proposed ACS), 2811, 2837, and 3086; Discussion on Bills Relating to Solid Waste Management and Disposal* (Trenton: New Jersey Legislature, 1997); Assembly Agriculture and Waste Management Committee, *June 12th Public Hearing Before Assembly Agriculture and Waste Management Committee: Assembly Bill no. 2627 ("Resource Recovery Facility Stranded Investment Cost Recovery Bond Act", Authorizing Bonds for $200 Million and Appropriates $5000)* (Trenton: New Jersey Legislature, 1997); Assembly Solid and Hazardous Waste Committee, *February 3rd Committee Meeting of Assembly Solid and Hazardous Waste Committee: Assembly Committee Substitutes for Assembly Bill nos. 515, 516, 517, 518, and 519: Issues Dealing with Solid Waste Management and Disposal* (Trenton: New Jersey Legislature, 1998); Assembly Solid and Hazardous Waste Committee,

February 26th Committee Meeting of Assembly Solid and Hazardous Waste Committee: Assembly Bill nos. 515, 516, 517, 518, and 519 (Proposed ACSs): Discussion on Proposed Committee Substitutes Related to Solid Waste Management and Disposal (Trenton: New Jersey Legislature, 1998); Assembly Solid and Hazardous Waste Committee *May 7th Committee Meeting of Assembly Solid and Hazardous Waste Committee: Assembly Bill nos. 515, 516, 517, 518, 519 (Proposed ACSs), and Assembly Bill nos. 1605 and 1889: Discussion on Proposed Committee Substitutes Relating to Solid Waste Management and Disposal* (Trenton: New Jersey Legislature, 1998); Assembly Solid and Hazardous Waste Committee, *June 4th Committee Meeting of Assembly Solid and Hazardous Waste Committee: Assembly Bill nos. 515, 516, 517, 518, and 519 (ACSs) and 1167: Discussion on Proposed Committee Substitutes Relating to Solid Waste Management and Disposal* (Trenton: New Jersey Legislature, 1998); Assembly Solid and Hazardous Waste Committee, *June 4th Public Hearing before Assembly Solid and Hazardous Waste Committee: Assembly Bill no. 1167 (Revises the Resource Recovery and Solid Waste Disposal Facility Bond Act of 1985, Appropriates $5,000)* (Trenton: New Jersey Legislature, 1998); Assembly Appropriations Committee, *July 27th Public Hearing before Assembly Appropriations Committee: Assembly Bill no. 2295 (Revises the Natural Resources Bond Act of 1980 and Resource Recovery and Solid Waste Disposal Facility Bond Act of 1985, Appropriates $5,000)* (Trenton: New Jersey Legislature, 1998); Jason P. Lien and Clinton J. Andrews, *Beyond Flow Control: Economic Incentives for Better Solid Waste Management in Mercer County, New Jersey* (New Brunswick, NJ: Rutgers University and the New Jersey Agricultural Experiment Station, 1998); and most succinctly Assembly Solid and Hazardous Waste Committee, *February 11th Public Hearing before Assembly Solid and Hazardous Waste Committee: Testimony from County Solid Waste Officials on the Evolution of County Solid Waste Management Activities Since the Waste Management v. Shinn, Carbone, and Atlantic Coast Decisions* (Trenton: New Jersey Legislature, 1999).

53 New Jersey Department of Environmental Protection, *Statewide Solid Waste Management Plan*, A-5.

54 New Jersey Department of Environmental Protection, *Statewide Solid Waste Management Plan*, 1.

55 Based on Larry Gindoff, interview by Jordan P. Howell, April 6, 2018 and Gary Sondermeyer, interview by Jordan P. Howell, May 2, 2018.

56 New Jersey Department of Environmental Protection, *Statewide Solid Waste Management Plan*, A-5–A-20.

57 Senate Environment Committee, *September 19 Committee Meeting*, 79.

58 The creation of a new, special-purpose *authority* was one tactic New Jersey counties would use to keep the debt for waste management separate from the county's general obligation debts. Usually, authorities like the PCFA received some type of financing guarantee from their home county in order to secure a better rate, which, while lowering borrowing costs, also put home counties on the hook for the debt.

59 Assembly Solid and Hazardous Waste Committee, *February 11th Public Hearing*, 11–14.

60 Assembly Solid and Hazardous Waste Committee, *February 11th Public Hearing*, 23–24.

61 New Jersey Department of Environmental Protection, 2006; p. A-2.

62 Senate Environment Committee, *September 19 Committee Meeting*, 26.

63 Assembly Solid and Hazardous Waste Committee, *February 11th Public Hearing*, 2–3.

64 Assembly Solid and Hazardous Waste Committee, *February 11th Public Hearing*, 3–6.

65 Assembly Solid and Hazardous Waste Committee, *February 11th Public Hearing*, 7.

66 Assembly Solid and Hazardous Waste Committee, *February 11th Public Hearing*, 89–91.

67 New Jersey Department of Environmental Protection, *Statewide Solid Waste Management Plan*, F-1.

68 New Jersey Department of Environmental Protection, *Statewide Solid Waste Management Plan*, table F-1.

69 New Jersey Department of Environmental Protection, *Statewide Solid Waste Management Plan*, table F-1.

70 Division of Solid Waste Management, *Draft New Jersey State Solid Waste Management Plan*, main section, 112.

71 Senate Environment Committee, *September 19 Committee Meeting*, 5.

72 New Jersey Department of Environmental Protection, *Statewide Solid Waste Management Plan*, B-6.

73 The 50 percent target for municipal solid waste is distinct from the famous 60 percent target established in decades past. The 60 percent target has applied to the "total waste stream" of all materials generated in New Jersey, including construction and demolition wastes, bulky items (like white goods and junk automobiles). If these heavy items are included in the calculations—as the NJDEP frequently does include them—then New Jersey has in fact met the 60 percent goal several times over the decades. But for the average resident or commercial user of the state's solid waste infrastructure, the 50 percent recycling rate target has never been met.

74 Gary Sondermeyer, interview by Jordan P. Howell, March 17, 2017.

75 Sondermeyer, interview by Jordan P. Howell, March 17, 2017.

76 For electronic waste, see, for example, Senate Environment Committee, *August 18th Public Hearing before Senate Environment Committee: Senate Committee Substitute for Bill no. 1861 (Electronic Waste Producer Responsibility Act); Senate Committee Substitute for Bill no. 2578 (Preservation of Landfill Space Act); Senate Committee Substitute for Bill no. 2615 (Recycling Enhancement Act)* (Trenton: New Jersey Legislature, 2005); Senate Environment Committee, *February 8th Committee Meeting of Senate Environment Committee; The Committee Will Take Testimony from the Public on How to Structure an Electronic Waste Management Program in New Jersey* (Trenton: New Jersey Legislature, 2007); and Senate Environment and Energy Committee *July 20th Committee Meeting of SENATE ENVIRONMENT AND ENERGY COMMITTEE "The Committee Will Hear Testimony from the Public on Potential Revisions to the Electronic Waste Management Act" Senate Bill No. 2973 Revises Electronic Waste Recycling Laws* (Trenton: New Jersey Legislature, 2015). For food waste and composting, see for example, Bridget Neary and Jordan P. Howell, "Organic Wastes Management in the Most Densely Populated State" (white paper, Rowan University, 2015), https://earth.rowan.edu/departments/geography/research/RSWL/_docs/neary_organics_nj.pdf; New Jersey Department of Environmental Protection, "Draft

Food Waste Reduction Plan" (2017), https://www.nj.gov/dep/dshw
/foodwasteplan.pdf.

77 See, for example, New Jersey Department of Environmental Protection, *Compliance Advisory Enforcement Alert: DEP Reminds Solid Waste Facilities of their Obligation to Plan for Transportation Emergencies* (Trenton: New Jersey Department of Environmental Protection, 2012); New Jersey Department of Environmental Protection, *Compliance Advisory Enforcement Alert: Illegal Dumping of Solid Waste* (Trenton: New Jersey Department of Environmental Protection, 2012); New Jersey Department of Environmental Protection, *Re: Temporary Solid Waste Equipment Authorization* (Trenton: New Jersey Department of Environmental Protection, 2012); and especially, New Jersey Department of Environmental Protection, *Guidance Document: Dealing with Increased Waste Generation in the Aftermath of Hurricane Sandy* (Trenton: New Jersey Department of Environmental Protection, 2012).

Chapter 6　Conclusions and Looking Forward

1 Joan Buehler, "Stricter Recycling Rules Hitting Towns," *The Retrospect* 118, no. 9 (2019): 1, 20.

2 Buehler, "Stricter Recycling Rules Hitting Towns," 20.

3 Prices reported in Tom Johnson, "Amid Crisis, Lawmakers Take Fresh Look at NJ's Recycling System," *NJ Spotlight*, August 16, 2019, https://www.njspotlight.com /stories/19/08/15/amid-crisis-lawmakers-take-fresh-look-at-njs-recycling-system/.

4 Catalina Jaramillo, "Incinerators in Camden, Chester among Nation's Most Polluting, Report Finds," WHYY.org, May 23, 2019, https://whyy.org/articles/incinerators-in -camden-chester-are-among-the-nations-most-polluting-report-finds/.

5 Per Covanta CEO Steven Jones; cf. "Covanta Holding Corporation (CVA) CEO Steve Jones on Q2 2018 Results - Earnings Call Transcript," Seeking Alpha, July 27, 2018, https://seekingalpha.com/article/4191347-covanta-holding-corporation-cva -ceo-steve-jones-q2-2018-results-earnings-call-transcript?part=single.

6 Charlie Kratovil, "As County Piles Garbage Higher, More Residents Say Middlesex Stinks," *New Brunswick Today*, January 19, 2019, https://newbrunswicktoday .com/article/county-piles-garbage-higher-more-residents-say-middlesex-stinks.

7 Michael S. Warren, "Mob-tied Illegal Dumping in N.J. May Finally End After Years of Warning," *NJ.com*, June 26, 2019, https://www.nj.com/news/2019/06 /mob-tied-illegal-dumping-in-nj-may-finally-end-after-years-of-warning.html.

8 New Jersey State Commission of Investigation, *Dirty Dirt: The Corrupt Recycling of Contaminated Soil and Debris* (Trenton: New Jersey State Commission of Investigation, 2016); and also New Jersey State Commission of Investigation, *Dirty Dirt II: Bogus Recycling of Tainted Soil and Debris* (Trenton: New Jersey State Commission of Investigation, 2019).

9 NJ State Commission of Investigation, *Dirty Dirt II*, 4.

10 Rina Li, "New Jersey Landfill under Fire for Mid-Inspection Sewage Violation," *WasteDive*, January 30, 2019, https://www.wastedive.com/news/new-jersey -landfill-sewage-violation-waste-management/547189/.

11 The town itself had not been affiliated with operating the landfill site since the arrival of the Meadowlands Commission nearly thirty years earlier.

12 Li, "New Jersey Landfill under Fire."

13 Rina Li and Leia Larsen, "New Jersey Supreme Court Orders Temporary Closure of Embattled Landfill," *WasteDive*, June 13, 2019, https://www.wastedive.com /news/judge-keegan-landfill-kearny-njsea/556036/.

14 Order issued by Jeffrey R. Jablonski in the matter of *Town of Kearny v. New Jersey Sports and Exposition Authority and New Jersey Department of Environmental Protection*, May 24, 2019, https://hudsoncountyview.com/wp-content/uploads /2019/05/Keegan-Landfill-order.pdf.

15 Atlantic County Utilities Authority, "New Jersey Solid Waste Disposal Fees 2018" (2019), accessed 22 August 2019, http://www.acua.com/uploadedFiles/Site /Disposal_And_Recycling/Location_and_Landfill/State_Tip_Fees.pdf.

16 Environmental Research and Education Foundation, "Analysis of MSW Landfill Tipping Fees, April 2017," April 2017, https://erefdn.org/wp-content/uploads /2017/12/EREF-MSWLF-Tip-Fees-2017.pdf.

17 Cf. Jordan P. Howell, Katherine Schmidt, Brooke Iacone, Giavanni Rizzo, and Christina Parrilla, "New Jersey's Solid Waste and Recycling Tonnage Data: Retrospect and Prospect," *Heliyon* 5, no. 8 (2019): https://doi.org/10.1016/j.heliyon .2019.e02313; and see also accompanying Jordan P. Howell, Katherine Schmidt, Brooke Iacone, Giavanni Rizzo, and Christina Parrilla, 2019 "New Jersey's Waste and Recycling Data: 1993–2016," dataset and supplementary materials published to Mendeley Data, August 10, 2019, http://dx.doi.org/10.17632/7yc4c3pmtp.2.

18 Of course this method is just an illustration. The tonnage figure includes construction and demolition debris in addition to *regular* solid waste, and the next-highest tip fee may not necessarily be the one that would be chosen by haulers, as economic rationality might dictate they choose the lowest possible rate that is feasible for their operations. As mentioned earlier tip fees vary across the state, as they do across the other comparison states. But the central point remains the same: state support of disposal markets in New Jersey could unlock significant value.

19 Sweden importing UK and Norway trash is the most famous recent example, though there are others. See Hazel Sheffield, "Sweden's Recycling Is So Revolutionary, the Country Has Run Out of Rubbish," *Independent*, 8 December 8, 2016, https://www.independent.co.uk/environment/sweden-s-recycling-is-so -revolutionary-the-country-has-run-out-of-rubbish-a7462976.html.

20 In this book I do not play the parlor game of speculating as to which towns or locations might be best suited to hosting new WTE and anaerobic facilities. Instead, I would point out that there is no shortage of vacant industrial land—in cities, suburbs, and exurbs—in New Jersey that could be realistically repurposed for such uses.

21 Thomas Marturano, interview by Jordan P. Howell, April 5, 2018.

22 See Jordan P. Howell, "Technology and Place: A Geography of Waste-to-Energy in the United States" (PhD diss., Michigan State University, 2013); also Jordan P. Howell, "Considering Waste-to-Energy Facilities in the US" (working paper, Rowan Solid Waste Lab, Rowan University, 2014), http://www.rowan.edu /colleges/chss/departments/geography/rswl/resources/whitepapers.html.

23 See State of New Jersey, *Draft 2019 New Jersey Energy Master Plan: Policy Vision to 2050* (Trenton: State of New Jersey, 2019), https://nj.gov/emp/pdf/Draft%20 2019%20EMP%20Final.pdf.

24 Cf. Bridget Neary and Jordan P. Howell, "Organic Wastes Management in the Most Densely Populated State" (white paper, Rowan University, 2015),

https://earth.rowan.edu/departments/geography/research/RSWL/_docs/neary
_organics_nj.pdf.

25 "Home Page," Terracycle, accessed September 7, 2019, https://www.terracycle.com
/en-US/.

26 Cf. Josh Lepawsky and Chris McNabb, "Mapping International Flows of
Electronic Waste," *Canadian Geographer* 54, no. 2 (2010): 177–195; Josh Lepawsky
and Mostaem Billah, "Making Chains that (Un)Make Things: Waste-Value
Relations and the Bangladeshi Rubbish Electronics Industry," *Geografiska
Annaler Series B: Human Geography* 93, no. 2 (2011): 121–139; Josh Lepawsky and
Charles Mather, "From Beginnings and Endings to Boundaries and Edges:
Rethinking Circulation and Exchange through Electronic Waste" *Area* 43, no. 3
(2011): 242–249; Josh Lepawsky, *Reassembling Rubbish: Worlding Electronic
Waste* (Cambridge, MA: The MIT Press, 2018).

27 See Mahbubur Meenar, Jordan P. Howell, and Jason Hachadorian, "Economic,
Ecological, and Equity Dimensions of Brownfield Redevelopment Plans for Environmental Justice Communities in the USA" *Local Environment* 24 (2019): 901–915.

28 See "Kalundborg Symbiosis," SymbiosisCenter Denmark, accessed August 22,
2019, http://www.symbiosis.dk/en/.

29 Cf. "Veolia and Microgrids: Advancing Power and Resiliency to Trenton, NJ,"
Veolia North America, accessed August 24, 2019, http://blog.veolianorthamerica
.com/veolia-and-trenton-new-jersey-resilient-microgrid; and also "CCMUA to
Pioneer First of its Kind Financing for Waste-to-Energy Microgrid in City of
Camden," Camden County, accessed August 24, 2019, http://www.camdencounty
.com/ccmua-to-pioneer-first-of-its-kind-financing-for-waste-to-energy-microgrid
-in-city-of-camden/.

30 The best example of this line of research (also, summarizing many other studies on
similar themes), is Samantha MacBride, *Recycling Reconsidered: The Present
Failure and Future Promise of Environmental Action in the United States, Urban
and Industrial Environments* (Cambridge, MA: The MIT Press, 2011).

31 Cf. Jordan P. Howell, "An Historical Geography of Michigan's Electricity
Landscape" (master's thesis, Michigan State University, 2010); also Jordan P.
Howell, "Powering 'Progress': Regulation and the Development of Michigan's
Electricity Landscape," *Annals of the Association of American Geographers* 101,
no. 4 (2011): 962–970;

32 See Finn Arne Jørgensen, *Making a Green Machine: The Infrastructure of Beverage
Container Recycling* (New Brunswick, NJ: Rutgers University Press, 2011).

33 There should be no doubt that the consolidation of major landfill operators
with major and minor recycling firms is one outcome of this cyclicality—
recycling firms become comparatively cheap over time, and landfill operators
see recycling services as a useful add-on to their main business line of hauling and
disposing at proprietary landfills. Investor returns on vertically integrated waste
service companies (like Waste Management, Republic, etc.) are uniformly strong
because the main business line—landfilling—is remarkably consistent and able to
mask the fluctuations inherent to current models found in the recycling industry.

34 For example, Matthias Herskind, "3 Ways to Profitably Reform the Recycling
Industry," *WasteDive*, October 9, 2018, https://www.wastedive.com/news
/recycling-industry-reform-/539058/.

35 Herskind, "3 Ways to Profitably Reform the Recycling Industry."

36 "Home Page," NOREXECO, accessed September 6, 2019, https://www
.norexeco.com.

37 Of course, there is frequently considerable money to be made by speculating in
commodity futures though the NOREXECO markets do not seem prone to that
for the time being.

38 See William Cronon, *Nature's Metropolis: Chicago and the Great West* (New York:
W. W. Norton, 1991).

Index

About the Author

JORDAN P. HOWELL is associate professor of sustainable business in the Rohrer College of Business at Rowan University. His work examines the human dimensions of environmental problems, with the intention of understanding the history of an issue in order to devise practical and meaningful solutions. He lives in New Jersey with his wife and daughter.